AIRLINER Icons Revealed

Welcome to this collection on airliner icons where we look at 15 aircraft that have changed the face of commercial aviation between the 1930s and 2026, as well as three others that may do so this decade.

There is no definitive poll that voted for the airliners chosen so I asked my aviation colleagues for their feedback and the consensus mostly included those you will read about in the coming pages. It is an eclectic group of airliners produced in Europe, the US and the USSR that many of you would have flown on, even if a 737 and A320 of today looks different from the first ones produced decades ago.

We live in fortunate times in that many of these aircraft have been faithfully restored by people who actually flew or maintained them and I urge you to get out and visit those museums before age gets the better of those experts.

Our collection starts in the 1930s with the ubiquitous Douglas DC-3 that introduced the idea of mass airplane travel and served so valiantly in World War Two, with more than 16,000 built in the US, Soviet Union and Japan. That's followed by a look into the Lockheed Constellation that Pan Am used to launch the world's first scheduled around-the-world service in 1947.

There is often a debate about which aircraft launched the Jet Age, but it was the UK-built Comet, although its accidents cast a shadow that lingers today. Comet's downfall is a heartbreaking story of human loss and how issues of metal fatigue were largely unknown and went undetected.

The stylish Boeing 707 certainly launched the commercial impetus behind jet travel and Boeing followed that with the 737 that arrived in 1957 and is still spawning new offspring today, while Concorde and the 747 certainly must rank in the highest echelons of airliner icons.

The trijet era filled a gap between narrowbody aircraft like the 737 and the 747 widebody and the success of the Airbus A300 and A320 series set the foundation for what the company is achieving today. The A380 also made the list, as did the reliable 777 that has been a steady performer through the recent difficult times at Boeing.

There is plenty more and I hope you enjoy learning all about how these aircraft evolved and what has made them successful. Writing and researching these airliner icons has been stimulating and inspirational about the future of our great aerospace industry.

Michael Doran
Editor

▼ **Qantas 747s made long-haul travel fun.** Qantas

United Airlines

Contents

Emirates

Herget Josef/Shutterstock

Qantas

ISBN: 9781836322269
Editor: Michael Doran
Senior editor, specials: Roger Mortimer
Email: roger.mortimer@keypublishing.com
Cover Design: Steve Donovan
Design: SJmagic DESIGN SERVICES, India
Advertising Sales Manager: Sam Clark
Email: sam.clark@keypublishing.com
Tel: 01780 755131
Advertising Production: Becky Antoniades
Email: Rebecca.antoniades@keypublishing.com

SUBSCRIPTION/MAIL ORDER
Key Publishing Ltd, PO Box 300, Stamford, Lincs, PE9 1NA
Tel: 01780 480404
Subscriptions email: subs@keypublishing.com
Mail Order email: orders@keypublishing.com
Website: www.keypublishing.com/shop

PUBLISHING
Group CEO: Adrian Cox
Publisher: Steve O'Hara

Published by
Key Publishing Ltd, PO Box 100, Stamford, Lincs, PE9 1XQ
Tel: 01780 755131
Website: www.keypublishing.com

PRINTING
Precision Colour Printing Ltd, Haldane, Halesfield 1, Telford, Shropshire. TF7 4QQ

DISTRIBUTION
Frontline Distribution Solutions Ltd, 2 Poultry Avenue, London, EC1A 9PU
Enquiries: 02074 294000 or info@flgroup.co.uk

Airliners Changing Aviation

Prepare yourself for a journey that starts in 1935 as we discover the aircraft that have altered the course of commercial aviation.

▲ **Air France Concorde with gear extended.** Herget Josef/Shutterstock

▶ **Douglas DC-3 in American Airlines livery.** Aircraft Images/Shutterstock

▶ **Emirates operated nearly half of all A380s.** Emirates

The world's first scheduled airline passenger service was operated on January 1, 1914, by St Petersburg-Tampa Airboat Line in the US. The aircraft used was a Benoist XIV flying boat that took 23 minutes to cross Tampa Bay compared with a multi-hour ferry trip.

After World War One, aircraft and pilots were easier to find and KLM, now the world's oldest airline operating under its own name, was founded in 1919. Many more soon followed and by the 1930s many national airlines had formed operating aircraft such as the Airco DH.4/DH.9, de Havilland DH.18, Breguet 14 and Junkers F.13.

As the industry grew it needed higher capacity and performance aircraft, which brings us neatly to where this collection of airliner icons kicks off, starting with the Douglas DC-3 that first flew in 1936. The DC-3 gave airlines the tools they needed to entice people off trains and into aircraft and did it far better than the Boeing 247 it competed with.

The collection starts in 1936 with the Douglas DC-3 that introduced mass air travel and cut fares by 50% before moving on to the Lockheed

Constellation, which came into civilian service in February 1946 on a Trans World Airline flight from New York to Paris, introducing the first pressurised airliner in widespread use.

The Jet Age came with the UK-built Comet in 1952, but a series of fatal crashes set that programme back years and left the space for Boeing to usher in the commercialisation of jet aircraft with the 707, which did so much to restore confidence in jet propelled airliners.

From there, Boeing moved quickly to develop and introduce the 737 in 1957, while France and the UK partnered to build Concorde, the world's first supersonic passenger jet airliner that entered service in 1976. Equally exciting was the arrival of Boeing's 747 in 1970 that made long-haul flying more affordable and comfortable for the masses.

The arrival of trijets like the DC-10 and Lockheed L1011 gave airlines more options to bridge the space between narrowbody jets and the cavernous 747, while the Soviet Union produced the Tupolev Tu-154 and Yakolev Yak-40 trijets that revolutionised regional flying in its vast territories.

Around that time, Airbus had been formed and introduced the A300/310 series in 1974, which was aimed at breaking the dominance of US manufacturers and rebuilding the European aerospace industry. For an encore, Airbus then launched the A320 Family in 1988, which in 2025 became the world's most delivered jetliner.

The torch then shines on aircraft that many readers would have flown on, such as Boeing's 777 series and the double-decker A380 that could seat up to 853 passengers. The 777 entered service in 1995, while the A380 started commercial service with Singapore Airlines in 2007.

Aviation is always on an upward path so the final section looks at three current aircraft that are changing the way we fly long-haul, ultra-long-haul and regional services. They are giving airlines new opportunities and significantly contributing towards lower carbon emissions, which means they will probably make it on the next version of *Airliner Icons*.

◀ **BOAC Comet 1A G-APAS.** Carlos Menendez CC BY-SA 2.0 via Creative Commons

◀ **Qantas 707 VH-EBA readying for departure from Seattle in 1959.** Qantas

▼ **As the launch customer of the truly iconic Boeing 747, Pan Am was able to influence many elements of the jumbo's final design.** Boeing

▲ **Basler BT-67 in Antarctica.**
Paul Helleman AAD

MASS Transport HERO

The Douglas DC-3 arrived at just the right time to propel passenger air travel and support Allied forces in World War Two.

In the early years of aviation, fledging carriers relied heavily on transporting cargo, which in many countries included mail delivery contracts. In 1931, US airlines earned around 83% of their revenue from carrying mail, but by 1940 that had switched to 70% coming from passengers.

Later in this publication you will read about the Jet Age, but long before that aviation entered its golden era of the 1930s, a time when aircraft and engine technology, along with competition for new military aircraft, propelled the industry forward at great pace.

In that environment the Douglas DC-3 stood head and shoulders above all others as it dramatically changed the aviation landscape, cutting US transcontinental journeys from nearly 40 hours with 13 stops to just 18 hours and three stops. It did that in style with sleeper cabins, unprecedented quietness, startling new levels of safety and operational costs below its competitors.

It was American Airlines' president, Cyrus Rowlett 'C R' Smith, who finally convinced aircraft manufacturer Donald Douglas Snr to build the new airliner by placing an order for 20 aircraft that saw the DC-3 become a reality. American Airlines (American) backed up the project by introducing its Admiral Clubs and chauffeur-driven limousines, innovative perks that most would think were dreamt up in more recent times.

DC-3 development

When Boeing went through its meltdown in the early 2020s the aviation world quickly realised that competition between the original equipment manufacturers (OEMs) was vital and an overdependence on one supplier was not a healthy outcome. In 1933, the first Boeing 247 entered service with United Airlines, but when Transcontinental

and Western Airlines (TWA) placed an order it was declined until the 60 ordered by United Airlines were delivered.

TWA turned to Douglas for a new airliner to compete with United, leading to a prototype DC-1 being developed that became the DC-2 in 1934 and laid the foundations for the DC-3. However, it was the dogged persistence of C R Smith that got the initially sceptical Donald Douglas over the line to launch the DC-3 and replace American's Curtis Condor II biplanes.

The order from American was finalised for ten 21-seat DC-3 passenger and ten Douglas Sleeper Transports (DST), which were essentially sleeper converted DC-2s. The DC-2 cabin was lengthened and widened from 66in to 92in to accommodate the 14 sleeping berths and a version with 21 seats was designated as the DC-3.

The DC-3 had a capacity for 21-32 passengers depending on the cabin layout or could carry 6,000lb of cargo, with a range of 1,500nm. It was originally powered by two Wright R-1820 Cyclone engines, but later civilian versions used Pratt & Whitney R-1830 Twin

◀ **Pan American Airways livery on DC-3 N877MG.** Ronnie Macdonald CC By 2.0 via Wikimedia

▲ **Douglas DC-3 Turkish Delight at airshow.** EvrenKalinback/Shutterstock

▶ **AIRVETERAN DC-3 at 2025 Kaunas Aviation Festival.** Adomas Daunoravicius/Shutterstock

▼ **DC-3 painted in Aer Lingus livery, 2017.** Miguel Mendez CC BY 2.0 via Wikimedia Commons

Wasp engines that gave the DC-3 a maximum speed of 233kts and a cruise speed of 183kts with a 23,200ft service ceiling.

The first DST flew on December 17, 1935, and entered service with American in June 1936, featuring 14 seats that converted to seven sleeper berths while a further seven berths were lowered from the ceiling. The seating-only version entered airline service with United Airlines in November 1936 and with 21 seats it became the standard version of the DC-3.

Entry into service

The significantly more comfortable, reliable and faster DC-3 was an overnight success for the Douglas Aircraft Company. The first aircraft entered service with American on June 25, 1936, with simultaneous flights from Newark in New Jersey and Chicago, Illinois, marking the start of the new era of mass air transport.

The major US carriers, including American, United, TWA, Eastern and Delta Air Lines, ordered more than 400 DC-3s and over the next two years more than 30 other airlines placed orders. Within two years of launch, the DC-3 carried 95% of all US airline traffic and by 1939 it was transporting 90% of all global passenger traffic.

US airline passenger numbers increased threefold in just four years after the DC-3 entered service and, just as is the case today, that surge in demand resulted in airfares dropping by approximately 50%.

Many decades before the arrival of low cost carriers (LCCs), the transformative nature of the DC-3

saw travellers ditch long-distance trains and flock to now-affordable air travel as their preference. By the mid-1940s, American's fleet consisted only of around 60 DC-3s and this standardisation around one aircraft type unwittingly set the model for today's LCCs, including Southwest, easyJet, Ryanair and many more.

International operators

The success of the DC-3 with US airlines did not go unnoticed across the rest of the world and shortly after its entry into service KLM Royal Dutch Airlines (KLM) was one of the first European carriers to add the aircraft to its fleet.

In 1936, KLM introduced the DC-3 on its Amsterdam-Batavia (now Jakarta)-Sydney service, which was then the world's longest airline route linking Europe with the Dutch East Indies and Australia.

In the late 1930s, airlines such as Air France, Sabena, Swissair and Aer Lingus were also early adopters of the DC-3 operating them on intra-European, Scandinavian and North African services. UK carrier Imperial Airways favoured flying boats and larger aircraft but British Overseas Airways Corporation (BOAC) operated large numbers of DC-3s on international routes during and after World War Two.

Further afield, Australia's flag carrier Qantas used the DC-3 for international connections to New Guinea and across the Pacific Islands to link overseas territories and mainline destinations. The DC-3 allowed Qantas, and Australia, to build international links in the region while using its larger aircraft for long-haul services.

The DC-3 also broke new ground in China with the China National Aviation Corporation (CNAC) acquiring the aircraft in the late 1930s. The rugged and reliable aircraft was ideal for the challenging environment and undeveloped airfields across China where the DC-3's short-field performance was crucial.

With similar conditions, Indian National Airways used DC-3s to expand international and regional

▲ **American Airlines DC-3 at Washington National, 1941.** Jack Delano Public Domain via Wikimedia Commons

▼ **Douglas DC-3 in American airlines livery.** Aircraft Images/Shutterstock

▲ **Movie star greeting off the DC-3 in 1938.** Harris Ewing Public Domain via Wikimedia Commons

services around the Indian subcontinent that allowed the fledgling nation to standardise air travel and effectively establish national connections.

In North America, Trans-Canada Air Lines, the forerunner to Air Canda, operated DC-3s for cross-border US-Canada services and on domestic trunk routes that fed into its international hubs.

In May 1945, Cubana de Aviación became the first Latin American airline to offer scheduled international flights from Havana in Cuba to Miami. The airline also operated DC-3s on some domestic routes well into the 1960s, a testament to the durability of the aircraft that was launched in 1936.

The attraction of the DC-3 to international airlines was driven by its economic viability, reliability and its availability soon after the end of World War Two. Commercial airline services were developing and the aircraft gave them the opportunity to be profitable from passenger services alone while its capability on rough airfields in all climates was a significant attraction.

Airline OEMs have to develop aircraft decades in advance of them entering service, but it is remarkable how often they get that balance right. In this case, Douglas produced an aircraft the world was crying out for, but without the persistence of American Airlines' Cyrus Rowlett Smith it might never have got off the drawing board.

It was also fortuitous that the huge demand for DC-3 aircraft during World War Two meant there were surplus aircraft everywhere after the war that made it much more affordable and simpler to start or expand an airline. In 2026, there are not enough new aircraft being built to satisfy demand and yet in 1939 Donald Douglas was able to get the DC-3 off the ground in under two years.

Military impact

Today, Airbus, Boeing and Embraer all run military and civilian aircraft production programmes in tandem, which often means a 737 or A330 is the same platform used under different designations in military service, such as the Boeing P-8 Poseidon and the Airbus MRTT multi-role tanker transport.

There were approximately 16,000 DC-3s produced, of which only 607, or less than 4%, were delivered to civilian operators during its production between 1926-1945. The other 15,000+ were delivered as military transports produced either in the US or under licence in the Soviet Union and Japan.

US factories in Santa Monica, Oklahoma City and Long Beach manufactured around 10,000 C-47 Skytrains while licensed builds in the Soviet Union added nearly 5,000 Lisunov Li-2s and almost 500 Japanese L2Ds.

The C-47 Skytrain first flew in 1941 and became the standard transport aircraft for the US Army Air Forces and many Allied air arms during World War Two. The US Navy operated its own version as the R4D and when in service with British and Commonwealth forces it was known as the Dakota. There was also a specialist type called the C-53 Skytrooper that was used for paratrooper missions.

▼▶ **Southwest Airlines DC-3 in San Francisco, 1948.** Bill Larkins CC BY-SA 2.0 via Wikimedia Commons

▼ **Chicago and Southern Airlines DC-3 in 1950.** RuthAS CC BY-SA 3.0 via Wikimedia Commons

▲ **Two-pilot cockpit on the DC-3.** TheFGG997 CC BY-SA 3.0 via Wikimedia Commons

The aircraft were primarily used for troops and cargo transport, paratrooper drops and towing gliders and played a pivotal part in airborne operations such as the D-Day landings in Normandy, France, during World War Two. Medical evacuations were another key mission with aircraft modified with space for stretchers and medical crews.

While the DC-3 was developed for airline passenger flights, its innate ruggedness and short-field performance made it ideal for military transport missions. The C-47s had strengthened undercarriages and floors, tie-down points and large cargo doors and were powered by 1,250hp Pratt & Whitney Twin Wasp R-1830 two-row 14-cylinder radial engines.

Post-war disposals

After the war, there were thousands of surviving C-47s spread around the world still in airworthy condition and available for civilian use. Many of these were converted back to DC-3s and used as airliners to form the backbone of passenger fleets for years to come, just as Cubana de Aviación did between the US and Cuba in 1945.

Along with this flood of DC-3s there were thousands of military pilots now looking to re-enter civilian life, so many of them got their first taste of

▼ **Douglas DC-3 Air Nostalgia in Australia.** Ryan Fletcher/Shutterstock

▲ **Air Atlantique DC-3 G-AMPY painted in RAF livery in 2008.** Rob Hodgkins CC BY-SA 2.0 via Wikimedia Commons

airline passenger services flying the DC-3. Unlike today, there were also thousands of trained mechanics to maintain the aircraft and globally airlines used these surpluses to spark their post-war revivals.

Flooding the market with more than 10,000 used C-147s all at once did nothing for sales of new aircraft and despite all its successes the civilian DC-3 went out of production in 1943 with the military versions produced until the war ended in 1945. The records show that only 607 DC3s were produced, but the number that actually operated with passenger airlines is much greater than that.

One last try

In an attempt to modernise its successful aircraft, Douglas launched a more powerful Super DC-3S with improved performance, comfort and operating economics. It was essentially a response to the newer types, like the Convair 240, while also providing existing operators with an upgrade path that reused existing airframes.

The DC-3S received a completely redesigned wing with improved aerodynamics that gave a higher critical Mach number, better high altitude performance and increased fuel capacity and range. It was fitted with Pratt & Whitney R-2000 Twin Wasp engines that gave higher cruise speeds and better climb, and upgraded landing gear and brakes.

The cabin was upgraded to hold up to 30 passengers with more modern seating and ventilation and a quieter environment. The DC-3S first flew on July 23, 1949, and gained certification in the early 1950s, but was a commercial failure. About 100 were produced or converted but almost all of those were delivered to military operators with only five going to airlines.

While the upgrades improved performance, the cost was close to the price of a competitor's new aircraft and the DC-3S lacked cabin pressurisation, which limited altitude and the capability to fly above bad weather. It was still based on an ageing 1930s airframe and customers preferred the Convair 240/340 rather than an upgraded pre-war aircraft.

After production ceased, DC-3s lived on for many decades in revenue service as passenger airliners, cargo aircraft and air charters, operating missions as

▼ **Douglas DC-3 N983DC painted as Air North.** Acroterion CC BY-SA 4.0 via Wikimedia Commons

diverse as skydiver shuttling, sightseeing, aerial spraying, military transport and missionary flying.

The Basler BT-67 is a heavily modified version of the DC-3 airframe that is produced in the US by Basler Turbo Conversions that was founded in 1990 to do these conversions. Converted BT-67s are used for operations in industries including cargo, military, cloudseeding and scientific research.

Conversions include fitting new Pratt & Whitney Canada engines, lengthening the fuselage, strengthening the airframe, upgrading avionics and making modifications to the wing leading edges and wingtips.

BT-67s have been vital on missions for the Australian Antarctic Division as their short take-off and landing capabilities and extended range and payload flexibility suit the harsh environment. The aircraft are fitted with retractable skis allowing take-offs and landings on snow or ice runways and other remote surfaces, making it operable in all conditions.

It is widely used for transporting personnel, supplies and equipment between bases on Antarctica and field camps and to conduct internal flights between the blue-ice runways that larger jets use to bring people to the continent and their deeper field locations.

The legacy

The DC-3 was not the first aircraft to carry passengers, but as the numbers show its arrival sparked the boom in passenger air travel, firstly in the US and soon after across the world. It also showed that airlines could make money without depending on delivering mail, which drew more investors and entrepreneurs into the industry.

The aircraft played a crucial role in World War Two; then those thousands of aircraft dispersed around the globe to launch new airlines in locations that previously had no air travel connections. Its presence cut US airfares by 50%, which opened the door to millions to ditch the train and hop on a DC-3… and that alone puts it in the airliner icons frame.

▲ **Douglas DC-8 AIRVETERAN at the airshow in Latvia in 2022.** Karolis Kavolelis/Shutterstock

▼ **Douglas C-47 military transport in 2019.** Ronnie Macdonald CC By 2.0 via Wikimedia

Sleek and Powerful Connie

It was the determination of Howard Hughes that got Lockheed to produce the classic Constellation.

A recurrent theme in this collection is that new aircraft launch on the back of emerging technology and refine that to become established and successful performers. The downside is that the next generation of technology is already being developed and as one aircraft reaches its zenith a new one comes along to knock it off its perch.

The Douglas DC-3 established the market for mass air transport and fought hard to stay at the peak, but as time marched on it lacked a key feature that we all now take for granted – pressurised cabins. Sometimes aircraft are so good at what they do that rival OEMs or airlines want to leapfrog the leaders, which is where the Lockheed Constellation enters the 'airliner icons' world.

Howard Hughes

After seeing how American Airlines' president Cyrus Rowlett 'C R' Smith pushed Donald Douglas to build the DC-3, the largest shareholder in Transcontinental and Western Air (TWA), the enigmatic Howard Hughes, started his own crusade with Lockheed for a new airliner.

Aside from being a wealthy industrialist and film producer, Hughes was a record-breaking pilot who in 1935 set a transcontinental speed record and, in 1938, he operated an around the world flight that demonstrated the possibilities of long-range air travel.

Hughes wanted an airliner that would give TWA a clear advantage over rivals flying the DC-3 so he personally pushed Lockheed to design an entirely new aircraft that could fly higher, faster and farther than any existing airliner. His unique combination of financial muscle, obsession with performance and sheer personal pushed Lockheed into creating the Constellation, a

▼ **Trans-Canada Air Lines' Constellation L-1049.** Minh Tran/ Shutterstock

◀ **Constellation cockpit at HARS Aviation Museum.** Peter Podouzek/ Shutterstock

▲ **TWA Constellation at New York JFK.** Leonard Zhukovsky/ Shutterstock

pressurised, long-range, high-speed airliner that redefined what airlines could offer passengers.

While Hughes initiated and largely financed the Constellation project, it was Lockheed that turned it into reality. To mitigate the risks for Lockheed, TWA placed an order for 40 aircraft and in 1939 work started on a new aircraft known internally as Model 049.

Connie takes shape

Lockheed had several non-negotiable targets for the Constellation that included pressurisation for passenger comfort at altitude, high cruise speeds and true transcontinental capability. The design was finalised in 1940 and the production of the first prototype began in California in 1941.

In specific terms, Hughes wanted an aircraft that could carry 40 passengers, cruise at 300kts at 22,000ft with a range of 3,000nm and, to reduce pilot fatigue, Lockheed developed hydraulically boosted flight controls, which was another first for the Constellation.

Designers adopted a low-drag fuselage and the distinctive triple-tail configuration that allowed the aircraft to fit into existing hangars while still providing sufficient vertical stabiliser area for control at high speeds. Power would come from four Wright R-3350 Duplex-Cyclone 18-cylinder radial engines with a large three-blade propeller that had a diameter of 15ft 22in.

The first flight of the Constellation prototype, registration NX25600, was on January 9, 1943 – delayed due to the onset of World War Two. In 1942, the US Army Air Force (USAAF) had taken control of the production line and Constellations, redesignated as the C-69, were used for long-range, high priority transport, including moving personnel and critical equipment across oceans.

In 1944, Hughes made a non-stop transcontinental flight from Burbank

▼ **First Qantas Constellation arrives in 1954.** Qantas

in California to Washington DC, in just over six hours with stops only for refuelling due to military restrictions. The demonstration flight was highly publicised and Hughes used it to highlight the Constellation's potential and reinforce TWA's claim that it would soon be operating the fastest and most advanced airliners in the world.

With priorities elsewhere, only 22 C-69s were built before the end of the war; seven of those never entered military service and were converted to civilian L-049s on the assembly line. The additional military funding helped Lockheed continue to develop the Constellation and in 1945 the aircraft successfully completed certification testing.

Entry into service

After the conflict ended, it was full speed ahead for Lockheed as it refined the aircraft into the L-049 and improved L-749 variant by adding more fuel capacity and reliability upgrades. Launch airline TWA received its first L-049 in November 1945, which operated a proving flight that departed from Washington DC on December 3, 1945, and arrived in Paris the following day after technical stops in Gander and Shannon.

The Constellation L-049/L-749 had a cruise speed of 260kts and a maximum speed of 291kts with a range of 2,500nm for the L-049 and the additional fuel capacity of the L-749 extended that to 4,300nm. The aircraft was 97ft 4in long, 24ft 6in high, had a wingspan of 123ft and could carry 44-60 passengers at a ceiling of 25,000ft with maximum take-off weights (MTOWs) of 102,000lb for the L-049 and 110,000lb for the L-749.

Lockheed began deliveries to airlines in October 1945 with initial deliveries to TWA and Pan Am and while international services took most of the limelight many were

▲ Airworthy Constellation at HARS Aviation Museum, Sydney. Peter Podouzek/ Shutterstock

▼ Thailand's prime minister leaves a Constellation in 1955. Daan Noske CC 0 via Wikimedia Commons

▲ **HARS Aviation Museum's Connie at the Avalon Airshow 2019.** Wirestock Creators/Shutterstock

▼ **Constellation HB-RSC landing in Samedan.** NAC CC BY-SA 3.0 via Wikimedia Commons

quickly introduced on domestic services as Constellations joined airline fleets.

Lockheed Martin's website shows the first Constellation began service on February 5,1946, when a TWA L-049, named the Star of Paris, flew a scheduled commercial flight from New York to Paris via Newfoundland and Ireland. As more aircraft arrived TWA used them on primary domestic trunk routes such as New York-Chicago and Los Angeles-Chicago.

Eastern Airlines and Capital Airlines also launched domestic services with Constellations in 1947 and early international airline operators included Air France, KLM, BOAC, Qantas, SAS and Avianca. Overall, more than 60 airlines operated the Connie throughout its service life with flag carriers such as Lufthansa, Iberia, Sabena, Air India, Royal Air Maroc, El Al and South African Airways adding aircraft.

It is unsurprising that many of these airlines operated long-haul transoceanic routes and with its pressurised cabin, high speed and passenger comforts the Constellation soon became the go-to aircraft for these demanding services.

Super Connie

Most new aircraft spawn variants that are either to improve performance or to meet the specific needs of an airline, such as extended range, alternative engine types or other modifications.

Across the family of Constellations the key features that made it so successful stayed the same, but as technology advanced it was added to the aircraft. The basic difference between the L-049 and the L-749 is the increased fuel capacity that extended the range from 2,500nm to 4,300nm to enhance its prized intercontinental capabilities.

With competition intensifying on long-haul routes, the next step came in 1951 when Lockheed introduced the L-1049 Super Constellation, which was designed to give airlines more seat capacity and improved performance and to reduce operational costs.

To do that, the fuselage was stretched from 97ft 4in to 113ft 7in to accommodate between 60-95 passengers, up from the 44-60 seats on the L-049, which immediately improved revenue for the airlines.

The performance gains came from switching to Wright turbo-compound R-3350 engines that recovered exhaust energy to increase effective horsepower from 2,200hp per engine on the original Connies to around 3,250hp on the Super. This boost gave the aircraft higher cruise speeds, better climb performance and improved high-altitude capability, all of which were important on long transoceanic routes.

The MTOW increased by 21% from 110,000lb on the L-749 to 133,000lb on the Super Constellation, and to structurally support that, the new version had strengthened wings and landing gear. While the wing kept its familiar shape it was redesigned internally to handle the extra payload and fuel capacity on the L-1049.

These improvements made the Super Constellation much better suited to dense intercontinental and domestic services and gave airlines more range and passenger capacity, evolving the platform into the most commercially successful Constellation type.

▲ **Vintage Lufthansa Constellation, Germany.** Joseph Creamer/Shutterstock

Squeezing the lemon

While these improvements optimised what piston engines could produce, the shadow of the Jet Age was just around the corner with the Comet and the 707 now under development. To squeeze everything it could out of the Constellation in 1957, Lockheed introduced the L-1649A Starliner that was designed specifically for maximum range whereas the Super Constellation was optimised for capacity and versatility.

To achieve that, the Starliner had a much longer span, high aspect-ratio wing that significantly increased fuel capacity and improved lift-to-drag efficiency that enabled ultra-long-range performance, with the span climbing from 123ft on the L-1049 to 150ft on the L-1649A.

With an increased range of up to 6,500nm and a service ceiling of 29,000ft, the Starliner was able to fly nonstop transatlantic sectors, albeit under favourable conditions with reduced passenger loads. Writing that reminds me of the early Air New Zealand 787 nonstop flights from New York to Auckland that unexpectedly had to refuel in Fiji when winds turned nasty.

The Starliner operated with the same Wright R-3350 engines and it derived its range advantage primarily from the new wing design and increased fuel capacity rather than by adding more power. In practice, the Super Constellation's stretched fuselage suited carrying more passengers

▼ **TWA Constellation at Pima Air Space Museum.** David Swierczek/Shutterstock

▲ **Air France Constellation departing Lydda, 1949.** David Eldan Public Domain via Wikimedia Commons

at shorter ranges on high density routes while the Starliner came into its own carrying fewer passengers on ultra-long-range services.

However, time was up for piston-engined and propeller-driven aircraft and just 44 Lockheed L-1649A Starliners were produced between 1957 and 1958. For some carriers it was a bridge between the new and the old engine technology and in a back-to-the-future moment, TWA was the largest operator with ten, followed by Air France and Lufthansa.

Qantas and Connie

The Starliner was the A350ULR of its day and the forerunner of the nonstop flights from Sydney to New York and London that Qantas is launching in 2027. Known as Project Sunrise, these specially modified A350s are sacrificing capacity for additional fuel tanks and, as a former Constellation operator, the Australian airline has solid history in that regard.

Qantas operated both L-749 and L-1049 Super Constellations with its first arriving in October 1947 and starting service from Sydney to London on December 1 that year. The trip took four days with overnight stops in Singapore and Cairo on the so-called Kangaroo Route, which was considerably quicker than the previous iteration that involved flying boats and early aircraft.

In 1954, Qantas introduced L-1049 Super Constellations that became the backbone of the airline's international services and the perfect showcase for what

▼ **Super Constellation at the Paris Air Show 2019.** Georges Segun CC BY-SA 3.0 via Wikimedia Commons

the Connie offered on long-haul routes. Constellations were used on services from Australia to London, Johannesburg (known as the Wallaby Route), Tokyo (Cherry Blossom Route) and San Francisco (Southern Cross Route) as well as many other ports along the Kangaroo Route including Singapore.

Farewell

By the late 1950s, Qantas began replacing its Super Constellations with seven Boeing 707 jets that started service in 1959 along with four Lockheed L-188C Electra turboprop aircraft. In May 1960, Super Constellation VH-EAG departed Sydney and ended the 16-year association between Qantas and the Constellation family that had served it so well.

That switch to jets was mirrored by airlines around the world starting with the Comet 1A, Douglas DC-8 and Boeing 707 that signalled the start of a new era, just as the Constellation started its own era in 1941. The Connie left an indelible mark on aviation by building a bridge between flying boats and wartime transports and jet airliners and played a pivotal role in the development of long-haul passenger networks globally.

Connie is without doubt a highly ranked airliner icon in its performance, speed and long-range capabilities, but also for its classic design that will remain stylish for decades to come. It has been with us for more than 80 years and is a tribute to the imagination and drive of Howard Hughes, who dared to dream bigger than anyone else of his time.

▲ **Breitling Super Constellation, landing in 2016.** Kambui CC BY 2.0 via Wikimedia Commons

◀ **Connie's iconic tail set-up is classic.** David Swierczek/ Shutterstock

Comet's Lost Opportunity

As the first jet passenger aircraft the Comet set the stage, but undetected and misunderstood failures put an end to its leadership.

Almost any conversation about the de Havilland DH106 Comet will focus on the fundamental design flaws, abysmal safety record and fleet groundings that ceded leadership of the Jet Age to US manufacturers Boeing and McDonnell Douglas.

Conversely, the Comet was the world's first commercial jet airliner and the first to operate regular jet-powered transatlantic flights, inaugurated by the British Overseas Airways Corporation (BOAC) in October 1958 from New York to London, albeit with a fuel stop in Gander, Canada, on westbound flights.

The aircraft ushered in unparalleled levels of passenger comfort and amenity and slashed travel times on routes operated by large turboprop aircraft, such as the McDonnell Douglas DC-6. BOAC's nine-stop services from London to Tokyo were cut to 36 hours on the Comet compared with more than 86 hours on its Argonaut (DC-4 variant) piston aircraft.

The word 'pioneer' is often used loosely, but de Havilland pushed many existing technologies to the limit in its quest to pioneer jet passenger services for the world. However, the company also subjected the Comet to a more rigorous testing programme than other contemporary commercial aircraft, but the devastating pressurisation and metal fatigue issues that later emerged were not fully understood at that time.

The second coming of the aircraft, the Comet 4, proved to be a reliable and effective aircraft globally that was in commercial service from October 1958 to November 1980. Including prototypes, 114 Comets were produced and of those 77 were the Comet 4B/4C variant, by far the most successful of all iterations of the jet.

An odd legacy

The failures of Comet 1 led to multiple and extensive accident investigations and inquiries that discovered the root causes of the accidents and enabled the Comet's return. These investigations developed techniques and analysis that are the foundation for today's accident response systems, such as aircraft reconstruction techniques, deep-sea salvage and the impacts of pressurisation and metal fatigue.

It is often said that the lessons learned from the Comet may have prevented emerging jet aircraft, such as the Boeing 707 and Douglas DC-8, suffering from similar undetected metal fatigue issues. Some of those learnings were incorporated in the thicker fuselage skin of Comet 4, along with fuselage reinforcements and demanding pressurisation testing that enabled the aircraft to be certified for 30,000 hours in operation, with one airframe tested to 100,000 hours.

▼ **Middle East Airlines Comet 4C at London Heathrow, 1963.** Ken Fielding CC BY-SA 3.0 via Wikimedia Commons

▲ **Dan-Air Comets at Gatwick, 1976.** MilborneOne CC BY-SA 3.0 via Wikimedia Commons

Aviation and aircraft development works best in a competitive environment and in that regard de Havilland grabbed the reins and introduced commercial jet aircraft that revolutionised air travel. Ultimately, it was that same competition that hastened the end of the Comet as it was swamped by faster, larger, longer-ranged and more cost-effective 707s and DC-8s.

Consolidation in the British aerospace industry saw de Havilland acquired by Hawker Siddeley in 1960 as orders for the Comet were dwindling. Strong competition was also coming from the Vickers VC-10, and with BOAC dropping the Comet from its transatlantic services the pioneering aircraft's troubled reign as the prime mover was coming to an end.

Many Comets changed hands and lived on with new operators led by UK charter airline Dan-Air, which operated the world's largest fleet of de Havilland Comets. Between 1966 and 1976, Dan-Air added 49 Comets to its fleet with the last aircraft retiring in 1980, although some of the aircraft had been obtained for spare parts and not for airline operations.

Operators that had switched to 707s and DC-8s were happy to sell their relatively young Comets to Dan-Air at a price significantly lower than the 737-200s and BAC One-Eleven 500s that the airline also considered. The Comets were a low-cost option to replace Dan-Air's ageing piston-engined aircraft, such

▼ **Comet 1XB at RAF Museum Cosford, UK.** DVLCOM/ Shutterstock

▲ **BOAC Comet 1A G-APAS.** Carlos Menendez CC BY-SA 2.0 via Creative Commons

as the Airspeed AS 57 Ambassador and the Avro York types that were nearing the end of their useful life.

Comet's early days

Despite coping with the immense pressure of fighting World War Two, the UK government was also concerned about the lack of British-built airliners and decided to form a committee to determine the country's future airliner needs. In 1942, it appointed aviation pioneer and former Minister of Aircraft Production, Lord Brabazon, to lead the effort.

The Brabazon Committee identified a variety of possible specifications, including one for a transatlantic pressurised aircraft that could carry one ton of payload at a cruising speed of 400mph. That concept caught the eye of Sir Geoffrey de Havilland, a fellow committee member and head of the de Havilland aerospace company.

In 1944, the UK Ministry of Supply issued tenders for five aircraft designed to meet different market needs. The successful aircraft types were for the production of the Brabazon, Airspeed Ambassador, Vickers Viscount, Britannia and Type IV, the world's first jet airliner ultimately known as the de Havilland DH106 Comet.

At that time, jet engines were in their early days and considered not yet reliable enough and too fuel hungry for commercial aircraft. However the committee, no doubt influenced by Sir Geoffrey de Havilland, accepted the company's proposal for a jet-propelled airliner and in 1945 the company received a development and production contract for the Type 106, which required de Havilland to design and develop both the airframe and engines.

The idea of one company designing the engine and the airframe seems fanciful today, but de Havilland was no stranger to jet propulsion through the development and production of the Vampire British jet fighter in the mid-1940s. The Vampire was the second jet fighter operated by the Royal Air Force and the first to powered by a single jet engine.

The engine, originally designated the Halford H-1, was the second British jet engine to fly, after Sir Frank Whittle's Power Jets W1, and the first to pass a type test and gain a type certificate for an aircraft propulsion turbine. The H-1, designed by Frank Halford and built by de Havilland, first flew in March 1943 in the Gloster Meteor and in September of the same year in the de Havilland Vampire.

Halford had previously worked in the de Havilland's design department and in 1944 de Havilland acquired his company and he rejoined as the chairman of de Havilland Engine Company. The H-1 became the de Havilland Goblin and a new H-2 design became the Ghost, with four Ghost 50 Mk1 engines powering the Comet 1.

Developing the aircraft

The first concept was for an airliner to fly the Atlantic with 1,000lb of mail and six passengers, but de Havilland believed a larger capacity airliner around the size of a Douglas DC-3 with four Ghost engines and carrying 24 passengers was the optimum design.

BOAC found the Type IV proposition so attractive that it agreed a deal to purchase 25 aircraft before the aircraft's first flight, although that was later amended to ten when the final contract was signed in December 1945. The following September, BOAC requested a larger 36-seat version and in December 1947 the redesigned jet was named the DH106 Comet.

With orders for 16 Comets in hand from BOAC, South American Airways and the British government, the project moved into production with the first Comet completed in 1949 and operating its maiden flight on July 27, 1949. It was displayed at the 1949 Farnborough Airshow registered as G-ALVG and exactly one year later the second Comet, G-ALZK, operated its first flight.

From the start of flight trials in July 1949, the first Comet completed 324 flying hours and by July 1951 these two Comets had completed more than 1,000 hours of development flying. The second aircraft (G-ALZK) started proving flights by the BOAC Comet Unit in April 1951 and over six

▼ **Dan-Air Comet 4B G-APMG at East Midlands,1975.** Rob Hodgkins CC BY-SA 2.0 via Creative Commons

▲ **Comets were often used for state visits.** OGL Wikimedia Commons

months logged more than 500 flying hours of crew training and route-proving operations.

By then, airlines around the world were showing interest and Australia's flag carrier Qantas sent a team of technical experts to observe how the Comet was performing. The aircraft was awarded type certification on January 22, 1952, and on May 2 that year, Comet G-ALYP departed London headed for Johannesburg as the world's first jetliner flight with fare-paying passengers.

Comet 1

The Comet 1 was longer but not as wide as a Boeing 737-100 and featured multiple design elements that set the aircraft apart from the competition. Its clean, low drag design included a swept-wing leading edge, integral wing fuel tanks, four-wheel bogie main undercarriage and the two pairs of Ghost MK 1 engines buried in the wings.

Embedding the engines was designed to avoid the drag of podded engines and allowed for a smaller fin and rudder since the hazards of asymmetric thrust were reduced. Large turboprop aircraft suffered with excessive cabin noise so to curb that the Comet engines were fitted with baffles to reduce noise emissions and extensive soundproofing was also implemented.

The cockpit was designed for two pilots, a flight engineer and a navigator, and was designed to be similar to the Lockheed Constellation cockpit to make pilot conversion training easier for key customers such as BOAC. The flight engineer controlled key systems such as fuel, air conditioning and electrical systems and the navigator occupied a dedicated station.

Other advanced features included full dual controls for the captain and first officer and irreversible powered flight controls. These controls increased safety by preventing aerodynamic forces from changing the directed positions and placement of the aircraft's control surfaces.

On the downside, armour had to be placed around the engine cells to contain debris from any serious failures and positioning the engines inside the wing made for a more complex wing design. However, having engines in the wing did reduce the risk of foreign object damage and the low-mounted engines and effective service panels simplified maintenance procedures.

In the first year, Comets carried 30,000 passengers and, with a break-even point below 50% passenger loads and reduced fuel burn, it was a commercial and marketing success. With a ceiling of 43,000ft and powerful Ghost engines, Comets flew well above poor weather that competitors had to go through while the higher speed and quieter cabin were added advantages.

In the summer of 1953, eight BOAC Comets departed London each week on Empire flights, with three to Johannesburg, two to Tokyo, two to Singapore and one to Colombo. It had given the UK a three to five-year head start over its competitors with Air France and Union Aéromaritime de Transport also operating the airliner on routes to the Middle East and West Africa.

Longer variants were planned with the Comet 2 attracting orders from Air India, British Commonwealth Pacific Airlines, Japan Airlines, Linea Aeropostal Venezolana and National Airlines, with a larger and longer-range Comet 3 being ordered by Pan Am. The Comet 2 made its first flight on August 27,

▼ **Comet with square windows.** Mike McBey CC BY 2.00 via Creative Commons

▼ **Olympic Airways Comet in Zurich.** Werner Friedli CC BY-SA 4.0 via Creative Commons

1953, and was planned to enter service in 1954… but stormy days were ahead for de Havilland and the Comet programme.

The accidents

The Comet was involved in 25 hull-loss accidents with 13 of those incidents resulting in fatalities, but it was five incidents in late 1952, 1953 and 1954 that saw the aircraft ultimately removed from service from April 1954 to October 1958.

In October 1952, BOAC flight 115 departing Rome failed to become airborne on take-off and ran off the runway into rough ground. Two passengers sustained minor injuries but the aircraft was written off. In March 1953, a Canadian Pacific Airlines Comet 1A failed to become airborne on take-off from Karachi and collided with an embankment, killing all five crew and six passengers.

For those who know about the 737 MAX losses it is disturbing to read that these two accidents were initially attributed to pilot error, but it was later found that the Comet's wing experienced a loss of lift at a high angle of attack. After those findings, de Havilland re-profiled the wings' leading edge and wing fences were added to control spanwise flow.

The Comet's safety record worsened considerably on May 2, 1953, when BOAC flight 783 crashed in a severe thundersquall six minutes after taking off from Calcutta in India, killing all 43 on board. Witnesses saw the wingless Comet on fire as it plunged into a village in a clear sign of structural failure.

The break-up began with a left elevator spar failure in the horizontal stabiliser and the inquiry concluded the aircraft had encountered extreme negative g-forces and that severe turbulence induced down-loading leading to the loss of the wings. Again the pilots were dragged into the findings but design changes also followed this inquiry.

On January 10, 1954, Comet G-ALYP broke up in mid-air over the Mediterranean while operating BOAC flight 781, killing all 35 on board. With no witnesses and incomplete evidence no obvious cause was found, although the investigating committee concluded that fire was the most likely cause.

BOAC voluntarily grounded its Comet fleet while the investigation continued and changes were made to protect the engines and wings from damage that might lead to fire. Sadly, with no apparent fault in the aircraft found, the British government decided

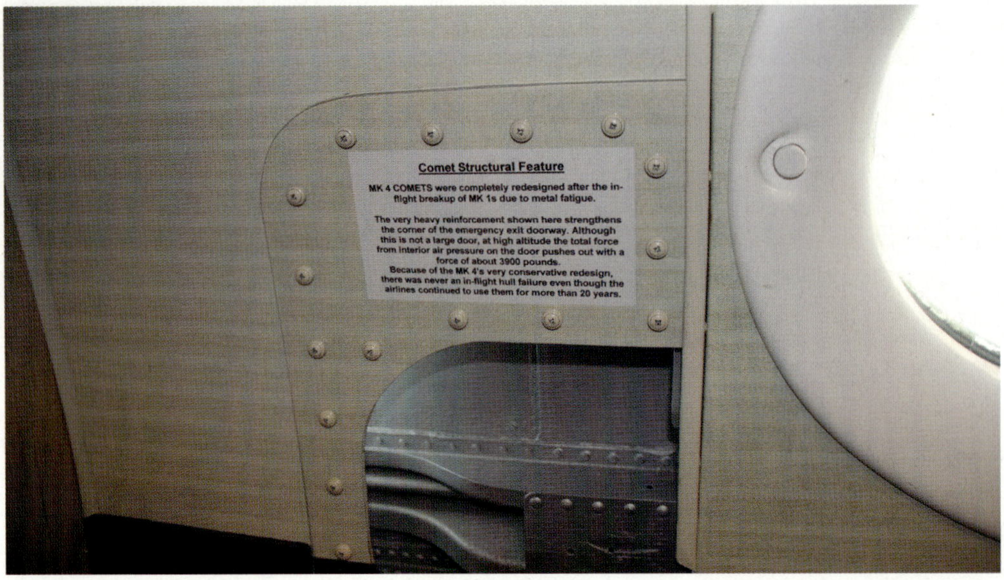

▲ **Cabin fix applied by de Havilland.** Clemens Vasters CC BY 2.0 via Wikimedia Commons

against a further public inquiry into the accident and Comet flights resumed on March 23, 1954. Just a few weeks later, on April 8, a Comet on charter to South African Airways operating a London-Johannesburg flight crashed into the Mediterranean while flying a sector from Rome to Cairo, killing all 21 passengers and crew. The Comet fleet was again grounded and after its Certificate of Airworthiness was revoked production was suspended and the BOAC Comet fleet was permanently grounded, cocooned and stored.

The Royal Navy was tasked to locate and retrieve the wreckage and on October 19, 1954, the Cohen Committee was established to examine the cause of the Comet crashes. The investigating team suspected metal fatigue as the most likely cause of the last two accidents and this was found to be correct.

The report said that fatigue crack growth from a countersunk rivet hole at the starboard aft corner of the rear Automatic Direction Finder (ADF) window had caused a catastrophic break-up of the aircraft in high altitude flight. Once the crack started the skin failed from the point of the ADF cut-out and propagated downward and rearward resulting in an explosive decompression.

Comet 4 redemption

After little more than a month, the Cohen inquiry closed on November 24, 1954, by finding that the basic design of the Comet was sound. In response, de Havilland began a refit programme to strengthen the fuselage and wing structure using thicker gauge skin and replacing the rectangular windows and panels with rounded versions.

All remaining Comets were withdrawn from service and all production Comet 2s were modified with thicker gauge skin to alleviate the fatigue problems, with most of those entering service with the Royal Air Force.

▼ **Comet 4 XS235, G-CDPA, ready for launch.** Alan Wilson CC BY-SA 2.0 via Creative Commons

BOAC remained loyal and ordered 19 Comet 4s in March 1955 and the upgraded aircraft first flew in April 1958, receiving its Certificate of Airworthiness on September 24 that year. The first Comet 4 was delivered to BOAC the next day and was used to inaugurate the London to New York services on October 4, 1958.

Compared with earlier versions, the Comet 4 had a longer range (2,800nm), higher cruising speed (450kts) and higher maximum take-off weight (156,000lbs), all made possible by the Rolls-Royce Avon engine with its increased thrust. It typically seated 74-81 passengers and up to 119 in a

special charter configuration and spawned type 4B and 4C variants.

It may be surprising to know that despite the earlier aircrafts' horrible safety record, the Comet 4 restored some credibility to de Havilland with 77 built, twice as many as all the other series combined.

It was in commercial service from October 1958 to November 1980, with airlines including BOAC, Kuwait Airways, Middle East Airlines, Qantas, Olympic Airways, Sudan Airways, Dan-Air, Pan Am/Mexicana, British European Airways, East African Airways, Air Ceylon and Malaysian Airways.

▲ **A very busy cockpit on the Comet 4C.** Geni CC BY-SA 3.0 via Creative Commons

◀ **Comet 1A G-ALYU failed fuselage panel.** Clemens Vasters CC BY 2.0 via Wikimedia Commons

UNITED STATES OF AM

Launching The Jet Age

It was Boeing's 707 that recovered passenger jet travel from the Comet's ashes and rebuilt the public's faith in air travel.

▲ **Modified 707 that was used as VC-137 for Air Force 1.** Grindstone Media Group/Shutterstock

The arrival of jet-propelled airliners completely changed the face of commercial aviation and opened the skies to millions of new travellers, particularly on long-haul transoceanic routes where journeys shrank from days to hours.

Termed the 'Jet Age', it ushered in faster, longer-range aircraft that flew higher to give passengers an experience not available on previous-generation turboprop aircraft, such as the Douglas DC-6 or Lockheed Constellation. Airlines were able to connect more cities and new routes than ever before opening opportunities for leisure, business and family reunion air travel.

Who launched the Jet Age?

The benefits of jet technology spawned complete families of new aircraft and paved the way for iconic people movers like the 747 and A380 that carried hundreds of people across continents in a fraction of the time. However, when the Jet Age is discussed, there is often a debate about which aircraft actually launched it, the Comet or the 707, with the answer perhaps depending on which side of the Atlantic the conversation takes place.

Objectively, it is clear that the British-built de Havilland Comet became the world's first commercial jetliner when it inaugurated service with BOAC from London to Johannesburg on May 2,1952. Boeing's 707 operated its first commercial passenger flight on October 26, 1958, on a Pan Am flight from New York to Paris – so on a quantitative basis, the Comet launched the Jet Age around six years before the 707 entered service.

However, due to its safety issues the Comet stumbled badly and forfeited its head start while the 707, with the support of the big US airlines, quickly flourished carrying four times as

▲ **Pan Am 707-321 N7249A at Dusseldorf Airport in 1969.** Peter Scharkowski/ Shutterstock

many passengers as the Comet to more destinations. Boeing may not have been the first to fly passenger jet airliners but the immediate success of the 707 stamped the Seattle giant as the leader of the new Jet Age, cemented further by the arrival of the 747 in 1970.

Early 707 days

Boeing built its early reputation manufacturing military aircraft such as the B-29 Superfortress, B-47 Stratojet and B-52 Stratocruiser, but was not as successful in civil aviation lagging behind Douglas, Lockheed and other rivals. Soon after World War Two it originated studies for a new jet transport that it believed would appeal to both military and civilian operators.

For military service, Boeing saw an opportunity to replace KC-97 Stratofreighters for aerial refuelling the United States Air Force's (USAF) new fleets of jet-powered fighters and bombers. In 1952, Boeing settled on developing a new quadjet prototype, the 367-80 (Dash 80) as a proof-of-concept aircraft for military and civilian operations.

The Dash 80 prototype had its first flight in July 1954 and the USAF used it as the platform for the KC-135 Stratotanker aerial refuelling and cargo aircraft. At the time, Boeing was highly dependent on military contracts for income so developing the Dash 80 was a significant risk for the company, particularly after losing US$15 million on the 377 Stratocruiser programme.

Boeing was keen to differentiate the new passenger jet from its 300-series propeller-driven fleet, so it designated the new jet fleet as the 700-series. The first in the series would be designated the 707, mainly because the marketing team thought that would be more appealing than naming it the 700.

Initially, Boeing offered the Dash 80/707 with a 132in-wide fuselage for four abreast 2-2 seating, but to compete with Douglas it widened it to 144in to allow five abreast seating. However, when Douglas launched the DC-8 with a fuselage width of 147in, airlines preferred the wider cabin so Boeing increased the width of the Dash 80 to 148in and six abreast seating.

The first civilian production type was the 707-120 (series 120) that had a longer, wider fuselage and greater wingspan than the prototype Dash 80 and could seat up to 189 passengers in a high density configuration. It was powered by four Pratt & Whitney JT3C-6 turbojets (civilian versions of the military J57) that produced 13,000lbf with water injection.

The 707-120 first flew in December 1957; it was designed for transcontinental routes and, like the Comet, often required a refuelling stop for flights across the Atlantic. As launch customers for the 707, American Airlines ordered 30 and Pan Am ordered 20, with the latter beginning 707 flights on October 26, 1958.

On January 25, 1959, American Airlines was the first domestic airline to fly its own jets, but was soon followed by Trans World Airlines (TWA) in March and Continental Airlines in June. TWA's then-president, the legendary Howard Hughes, saw the promise of the 707 early and ordered eight 707-120s in February 1956, which began service on the New York to San Francisco route.

707 variants multiply

Keen Boeing followers will know that it used customer numbers to identify specific options and livery specified by customers. The code consisted of two-digits affixed to the model number to identify the specific version, thus a 707-320B

▼ **Air Seychelles 707-324C on short final at Paris Orly.** Felix Goetting GFDL 1.2

sold to Pan Am had the model number 707-321B, with 21 being Pan Am's customer number.

Similarly, a shortened 707 variant produced for Qantas was designated 707-138 as 38 was the Qantas customer number. A total of 56 of the 707-120s were built, including seven short-bodied -138s with the last -120 delivered in May 1960. The Qantas 707s were shortened by 10ft to allow for full-load take-offs at a refuelling stop in Fiji and seven were delivered between June and September 1959.

Boeing had bet big on the 707 to turn around its financial fortunes and to do that it was extremely flexible in tailoring aircraft to a customer's needs. Between 1956 and 1978, Boeing produced 1,010 aircraft in various configurations of the 707, including passenger, cargo and military versions.

◀ **Boeing test pilots on the 707 in 1958.** Joseph Tymczszyn Wikimedia

Intercontinental 707-320 arrives

Following the launch of the 707-120, Boeing produced multiple civilian variants, with the main being the -120, -320, -420 series and the short-range 707-020 later known as the 720. Sub-variants were used to denote engine, airframe or other changes, such as the -120B, -138, -138B, -220, -320B and -320C.

The most popular type was the 707-320 series and with 580 produced it accounted for 67% of the civilian 707/720 production. The 707-320 Intercontinental is a stretched version of the 707-120 fitted with more efficient Pratt & Whitney JT4A turbojet engines that increased capacity to 189 passengers, up from 174 in the 707-120.

It carried more fuel in a longer wing increasing the range to 3,750nm and allowing the 707 to operate full transoceanic flights. The -320 operated its first flight on January 11, 1958, before Pan Am operated the first commercial flight in August 1959.

The 707-320B was upgraded with the more fuel efficient Pratt & Whitney JT3D turbofan engines and wing modifications that increased the wingspan by three feet. The new wing featured a second inboard kink, a dog-toothed leading edge and curved low drag wingtips and these changes enabled the range to increase to 5,800nm.

The 707-300B series could operate nonstop westbound flights from Europe to the West Coast of the US and from the US to Japan. The first 707-320Bs entered service in June

▼ **MEA Boeing 720 derived from the 707 platform.** Kambui CC-BY 2.0 via Creative Commons

▲ **Boeing 707 delivered to Qantas in 1968.** Qantas

▼ **African airlines were quick to see the promise of the 707.** Airlinephoto/ Shutterstock

1962 with Pan Am and after this the success of the 707 was assured on both domestic and international routes with it rapidly becoming the most popular jetliner of its time.

The most produced variant with more than 300 made was the 707-320C that had a convertible passenger-freight configuration. This model added a strengthened floor, a new cargo door and three section leading edge flaps to the -320B version. From 1963, the wing improvements were also added to the -320B and after modification these aircraft were known as the 707-320B Advanced.

While the configuration could be changed the majority of these types were delivered as passenger aircraft and with the addition of two new emergency exits, one on each side of the fuselage behind the wing, the maximum capacity increased to 219 passengers, with a typical one-class layout seating 194 and 141 seats in a two-class configuration.

Northwest 707 convertibles

In the US, Northwest Airlines (NWA) jumped at the opportunity to fly the 707-320 and in 1963 the airline revealed its revolutionary cabin design concept that would allow it to substantially increase transpacific cargo capability without reducing passenger capacity.

NWA added that the "large door cargo compartment, with space for five tons of cargo, was incorporated into the design (to) allow Northwest to provide, on the same flight, cargo lift equivalent to that provided today on the carrier's piston-engined all-cargo aircraft as well as the passenger capacity of NWA's present jet transports".

Northwest planned to operate a fleet of five -320B and 31 -320C types, with deliveries between1963 and 1968, with the -320Bs to be upgraded to the -320C configuration. The airline said some of the C models would be transformed into dedicated freighters with the 707s operating on Pacific and high-traffic US routes.

Rolls-Royce joins in

The 707-420 was almost identical to the -320 apart from the engines, which switched from Pratt & Whitney JT3D powerplants to the Rolls-Royce Conway 508 (RCo.12) turbofans. A look at the customers that pushed for the -420 suggests

the motivation to source Rolls-Royce engines ahead of the US types was heightened by support for European aerospace manufacturers.

Conway engines were used on aircraft such as the Handley Page Victor, Vickers VC10, Douglas DC-8 and the Boeing 707-420. The choice had Boeing approval with the OEM calculating the Conway engine would increase the -420's range by 8% ahead of the JT4A, although the actual operational gain turned out to be 2%.

With the 707 and DC-8 in hot competition, it is interesting to note that Conway engines were selected for both aircraft, with Trans-Canada Airlines, Alitalia and Canadian Pacific adding the Rolls-Royce engine to their DC-8s and BOAC, Lufthansa, Varig, El Al and Air India selecting it for their 707s.

▲ **Kuwait Airways 707-320C on approach to London Heathrow.** Steve Fitzgerald GFDL 1.2

◀ **Sudan Airways 707-3J8C departing Frankfurt Airport.** Felix Goetting GFDL 1.2

▼ **Air France was an early adopter of the 707.** Richard Vandervord CC BY-SA 4.0 via Wikimedia Commons

▲ **Air Zimbabwe 707-330B approaching London Gatwick.** Torsten Maiwald GFDL 1.2

▼ **Qantas operated 707s across the Pacific to warm receptions.** Qantas

The Conway was revolutionary as it was the first turbofan and the first commercial engine equipped with internally air-cooled turbine blades. This feature allowed for an extremely high exhaust temperature of 1,247°F (675°C), which partially accounted for the engine's high efficiency.

The downside was the addition of the Conway engines made -420 take-offs the loudest of any aircraft at airports and over communities until Concorde took that mantle. The 708-420 was built with a distinctive scalloped exhaust nozzle incorporating noise suppression and mechanical and aerodynamic thrust reversal that was developed and supplied by Rolls-Royce.

The first commercial flight of a 707-420 was operated by Lufthansa in March 1960, although the new type had to fight some bureaucratic battles along the way. The UK Air Registration Board initially refused to issue a certificate of airworthiness for the new version, citing insufficient yaw control, excessive rudder forces and the danger of over-rotation on take-off.

To overcome these concerns, Boeing added 40in to the vertical stabiliser, fitted an underfin and applied full instead of partial rudder boost to prevent over-rotation. Apart from the underfin, these modifications became standard on all variants and were retrofitted to earlier 707s.

The orphan 707

Perhaps the least well known of any Boeing airliner is the 720, an aircraft derived from the 707 that fills a gap in between the 737 and the 707. The aircraft was announced in July 1957 as the 707-020 to operate short to medium-range flights from shorter runways.

US carrier United Airlines was the launch customer and played a role in developing the aircraft, including the name change to the Boeing 720. It first flew in November 1959 and gained a type certificate in June 1960 before entering service with United on July 5, 1960.

It is derived from the 707-120 with four fewer frames in front of the wing and one fewer aft, reducing the overall length by 8ft 4in. It also had a modified wing structure and a lightened airframe that gave it a lower gross weight, higher top speed and improved take-off performance.

The 720 was powered by four Pratt & Whitney JT3C turbojets that gave it a range of 2,800nm carrying 131 passengers in a two-class layout. In 1961, the 720B entered service fitted with Pratt & Whitney JT3D turbofans, increasing range to 3,200nm with 156 passengers in a one-class cabin.

After building 154 aircraft, the 720 production line was closed in 1967, with 65 720-000s and 89 720-000B jets delivered. It was operated by more than 50 airlines globally, including American Airlines, Alaska Airlines, TWA, Air Ceylon, Continental, Braniff, Pan

▲ **Saha Airlines (Iran) Boeing 707-3J9 flight crew.** Danial Haghgoo GFDL 1.2

Am, Lufthansa United, Aer Lingus, Ethiopian, Pakistan International and Aerocondor Colombia.

707 legacy

There is no single aircraft that solely defined the Jet Age, but Boeing took jet technology and ran with the opportunity faster than its competitors. The 707 was a safe, reliable and very comfortable jet that introduced a whole new demographic to long-haul flying, priming the market demand for the high capacity 747 jumbo that entered service in 1970.

The 707 also mopped up the public's poor perception of jet aircraft safety following the Comet's disastrous crashes, while the quieter cabin, higher speed and elegance brought a luxury feel to the golden age of flying.

The aircraft trailblazed routes in all parts of the world and brought transoceanic travel within reach of many, which had previously been the preserve of more well-heeled travellers. The 707 became the backbone of fleets at airlines that lead the industry today, such as British Airways, American Airlines, United Airlines, Lufthansa, Qantas and Japan Airlines.

The aircraft capitalised on jet technology to take centre stage for decades and when new technology and market demand overtook the 707 it gracefully segued to the arrival of the 747. It is truly an 'airliner icon' and stands tall among the greats in the annals of aviation.

◀ **South African airline Interair's 707-323C registration ZS-IJI.** Bob Adams South Africa

Setting the Standard

Boeing went from following in the wake of its competitors to becoming the clear leader of narrowbody jets with the 737… and stayed there for decades.

After World War Two, Boeing was lagging behind its competition, particularly Lockheed and Douglas, in the civil sector and set about developing a new jet transport that it believed would appeal to both civilian and military operators.

In 1952, those studies led to Boeing's 367-80 (Dash 80) four-engined prototype and from that the launch of KC-135 Stratotanker and the civilian 707 series. The first 707-121 performed its initial flight in December 1957 and entered service with Pan Am in October 1958.

For all the reasons detailed in a previous chapter, the 707 was a roaring success for Boeing and introduced millions of people to the Jet Age. The aircraft was a step up on anything seen previously and paved the way for the introduction of the 727 trijet in 1964 and the twin-engined 737 in 1968.

A star is born

The successes of the 707 and 727 gave Boeing the confidence to develop a new aircraft to supplement the 727 on short and thin routes with research work indicating a market for

▼ **Malaysia Airlines has added 737-8s to the fleet.** Boeing

50-60 seat airliner flying routes of 50 to 100 miles.

Design work on the 737 started in 1964 although Boeing was late into this market and chasing new aircraft such as the BAC One-Eleven, Douglas DC-9 and Fokker F28. However, it soon brought the design to life by using 60% of the structure and systems of the existing 727, particularly the fuselage that previously had been derived from the 707.

Using the wide 707 fuselage gave Boeing an advantage in that the 737 could seat passengers six abreast while its competitors were offering five abreast seating, an advantage that airlines were quick to adopt. In 1965, launch customer Lufthansa ordered 21 737-100s and in April 1965 United Airlines placed an order for 40 of the larger capacity 737-200.

The initial 737-100 had its first flight on April 9, 1967, was certified on December 15 that year and entered service with Lufthansa in February 1968, making Lufthansa the first non-American airline to launch a new Boeing aircraft. The -100 version was soon overtaken by higher capacity variants and only 30 were produced, with Lufthansa the only significant operator of the type.

August 8, 1967, marked the first flight of the 737-200, with certification granted on December 21 and the inaugural United Airlines flight operated from Chicago to Grand Rapids in Michigan on April 28,

▲ **United Airlines has a fleet of nearly 600 Boeing 737s.** United Airlines

▲ **The cockpit of a 737-300, circa 1994.** Alasdair McLellan CC BY-SA 3.0 via Wikimedia Commons

1968. This variant seated 130 while the 737-100 had been increased at Lufthansa's request to 100.

The 737-200 Advanced came into service with All Nippon Airways in May 1971 and featured more powerful engines, longer range, improved aerodynamics and automatic wheel brakes, and could carry additional fuel. The 737-200 Advanced variant became the standard production model in June 1971.

In the lexicon of the 737, these two variants became known as the Originals or the First Generation types with 30 of the smaller -100s and 1,114 of the -200 series manufactured from 1967 until the last 737-200 was delivered to China's Xiamen Airlines in August 1988.

Classic era

The second generation of 737s – collectively known as the Classics – encompassed the 737-300, 737-400 and 737-500 types. The Classic series was produced between 1981 and 2000 and in total 1,988 were delivered in that time, with 1,113 of those being the 737-300 variant.

Based on the success of the 737-200, Boeing saw an opportunity to increase seat capacity with a longer-range aircraft while retaining commonality with the previous 737s. Development work started in 1979 and the 737-300 series first flew in February 1984, entering service with US Air in December of that year.

Boeing marketed the Classics near and far and was able to gain commitments from high quality carriers as launch customers for each of the series. The launch

▼ **Many airlines use 737s for promotions.** Sun Country Airlines

airlines all showed confidence with firm orders that secured their place in the queue and included Southwest and US Air for the -300, Piedmont for the -400 and Southwest again for the 737-500 aircraft.

The -300 was powered by the quieter and more fuel efficient CFM56-3 turbofan, but that posed an engineering challenge because of the low ground clearance of the 737 that was a characteristic of the 707 fuselage. The solution was to place the engine ahead of the wing and move the engine accessory gearbox to the side, giving the 737 its distinctive non-circular air intake.

The 737-300 is 109ft 7in long and can carry 149 passengers, with a range of 2,255nm; the stretched -400 is 119ft 7in long with a range of 2,060nm and 168 passengers; and the shrunk 737-500 is 101ft 9in long and can carry 132 passengers, with a range of 2,375nm.

Boeing launched the 737-400 in 1985 to fit in between the 737-300 and the 757-200 and compete more effectively with the Airbus A320 and McDonnell-Douglas MD-80. The -400 first flew in February 1988 and entered service with launch customer Piedmont Airlines on September 15, 1988; the last of the series was delivered to CSA Czech Airlines in February 2000.

The -400 series was a popular choice among US carriers with 489 delivered, but when United Airlines, previously a loyal Boeing customer, placed a significant order with Airbus for the A320 it was time for an updated aircraft. The -400 series morphed into the 737-800 that was first delivered in April 1998 and became the highest selling model in the family.

The last Classic to launch was the 737-500, a shorter and longer-range aircraft designed to replace the 737-200 that had gathered a broad spectrum of airline customers. It was launched in 1987 and first flew on June 30, 1989, with Southwest Airlines taking delivery of the initial -500 in February 1990.

The -500 was favoured by carriers who operated longer routes with fewer passengers because its operating economics were more favourable than using the larger 737-300 on those services. Boeing delivered 388 of the -500 series with the last delivered to All Nippon Airways in July 1999, although by then its replacement, the 737-600, had been in service for nearly a year.

Next generation

Airline deregulation in the US had seen a rapid proliferation of new carriers and routes and with fuel prices increasing, airlines were looking for more efficient narrowbodies. Facing increasing pressure from the Airbus A320, Boeing knew it needed to upgrade the 737. So in 1993 it announced the third derivative designated as the 737 Next Generation (NG).

The NG was not a superficial change as compared with the Classic the new series had a larger redesigned wing, longer range, increased fuel capacity, higher MTOW and new glass cockpit. It was powered by two CFM56-7B turbofan engines that reduced fuel consumption, carbon emissions and maintenance costs with an Enhanced engine introduced in 2015.

A Speed Trim System (STS) was introduced on the Classic series and was updated to include a stall identification function on the 737NG type. The updated STS operates at any speed on the 737NG, is triggered by an airspeed sensor and

▲ **Boeing's 737-300 is part of the Classic series.** Aero Icarus CC BY-SA 2.0 via Wikimedia Commons

▼ **Alaska Airlines 737-400 in Vancouver.** Makaristos Public Domain via Wikimedia Commons

▲ **Aeromexico operates more than 100 737s.** Makaristos Public Domain via Wikimedia Commons

commands Airplane Nose Down as the aircraft slows down.

The New Generation series encompassed the 737-600, -700, -800 and -900, which offered passenger capacity up to 130, 148, 184 and 215, respectively. The engine, wing and fuel capacity changes increased the 737's range by 900nm to more than 3,000nm opening up nonstop transcontinental routes for the narrowbody.

The NG programme was extensively tested using three -600s, four -700s and three -800s and the first two types operated their maiden flights in 1997, starting with the -700 on February 9 and the -800 on July 31. The -600 had its first flight on January 22, 1998, and the -900 was a late addition with a maiden flight on August 3, 2000.

The final commercial 737NG was delivered in January 2020 when two 737-800s were handed over to China Eastern Airlines, marking the end of a 23-year run that delivered 6,765 New Generation aircraft, not including several hundred military and private jets on the platform.

Nearly 5,000 of the deliveries were 737-800s and more than 1,100 were -700s – illustrating how popular these two variants were with airlines and passengers. At 129ft 6in long the 737-800 is nine feet longer than the -700 with the capacity to carry 189 passengers in a one-class cabin or 162 in a typical two-class layout.

For more than two decades it has been the workhorse of Boeing's fleet and proven itself with full-service and low-cost carriers all around the world. It replaced the 737-400 and was Boeing's competitor to the

▼ **Qantas 737-800 painted in a 1970s' livery.** Qantas

emerging A320, a battle that has carried on into the A320neo and 737 MAX era.

It also replaced the discontinued McDonnell Douglas MD-80 and MD-90 and for many airlines it was an ideal choice to take over from ageing Boeing 727 trijets. Just as easyJet built its business around the A320, LCCs Ryanair and Southwest have been equally loyal to the 737, with the trio each operating hundreds of their chosen narrowbody.

As Airbus gained ground, Boeing launched the 737-900ER, the final and largest version of the series in competition to the A321's success. The -900ER was launched in July 2005 and first flew in September 2006, entering service with Indonesian LCC Lion Air on April 27, 2007.

The addition of two auxiliary fuel tanks in the cargo hold and standard winglets increased the range to 2,950nm with a capacity for up to 220 passengers in a one-class configuration. Extending the range proved a commercial success with more than 500 of the -900ER sold, compared with just 52 of the standard -900 being delivered.

Time for MAX

In the early 2000s, Boeing was caught in a bind between developing a clean sheet narrowbody or adding new engines to the existing 737 platform with its links back to the 707 that entered service in 1968. When Airbus launched the A320neo Family in 2010 offering new engines that significantly reduced fuel burn and

◀ **Nauru Airlines operates 737s between Australia and its Pacific home.** Nauru Airlines

▼ **Boeing builds its 737 series in its spiritual Seattle home.** Boeing

▲ **Fiji Airways uses the MAX on flights to Australia.** Fiji Airways

emissions, the pressure ramped up on Boeing to make a decision.

That pressure increased when Airbus secured commitments for nearly 700 A320s at the 2011 Paris Air Show, but a bigger setback was soon to come when long time Boeing customer American Airlines placed an order for 460 narrowbodies, including 260 A320/A320neos and only 100 737NGs.

On August 11, 2011, Boeing announced it would launch a re-engined 737 MAX to compete with the A320neo and abandon the development of a clean sheet design. Boeing said fuel savings would be comparable with the A320neo and secured Southwest Airlines as the launch customer.

The MAX family initially consisted of the MAX 7, MAX 8 and MAX 9, which succeeded the most popular NG types, the 737-700, -800 and -900ER, respectively. The 737 MAX 10 resulted from customer requests for a larger aircraft to compete with the A321neo, which was easily outselling the MAX 9 variant.

The official type designations are the 737-7, 737-8, 737-9 and 737-10 although by the start of 2026 the MAX 7 and MAX 10 were mired in delays and yet to be certified. The first 737 MAX 8 completed manufacture in December 2015 and was soon in the air with its first flight on January 29, 2016.

The 737-8 gained FAA and EASA certification in March 2017 and was delivered to the then Malindo Air, a Malaysian full-service airline and a subsidiary of Indonesia's Lion Air Group now branded as Batik Air Malaysia. This first 737-8 entered service on May 22, 2017, and on August 29 that year, launch customer Southwest received its first 737 MAX.

The second aircraft off the line was the 737-9 – 15ft longer than the Max 8, which has a typical two-class layout of 193 seats with a range of 3,300nm, compared with the 737-8's two-class 178 seats and 3,500nm range. The 737-9 first flew on April 13, 2017, gained certification in February 2018 and was delivered to Lion Air on March 21 that year.

Tragedy awaits

No review of the 737 MAX can ignore the fatal crashes that saw the aircraft grounded globally nor the failure of an Alaska Airlines MAX 9 that developed a gaping hole in the fuselage while flying. However, this chapter has neither the room nor remit to dissect those events in detail other than their effects on the 737 MAX programme.

Fatal accidents involving a Lion Air 737-8 in October 2018 and an Ethiopian Airlines 737-9 in March 2019 led to the grounding of the entire 737 MAX fleet from March 2019 to November 2020. The accidents cost the lives of 346 passengers and crew and investigations revealed a faulty sensor sent incorrect data to the Maneuvering Characteristics Augmentation System (MCAS).

MCAS is a flight control software system designed to prevent aircraft stalls by automatically pushing the aircraft's nose down when an aircraft pitches up excessively. Investigations revealed that Boeing had not adequately disclosed the MCAS system to operators and the FAA had not adequately certified the system.

Following extensive software and hardware redesign and pilot training, the FAA cleared the 737 MAX to return to operation on November 18, 2020, and EASA did likewise in January 2021. Other nations progressively ungrounded the aircraft although the Civil Aviation Administration of China deliberated until December 2021 before giving it clearance to return.

▼ **Singapore Airlines once operated 737-800s.** Singapore Airlines

MAX unplugged

On January 5, 2024, Alaska Airlines flight 1282 was set to operate a flight from Portland in Oregon to Ontario, California, using a two-month-old Boeing 737-9. Shortly after take-off a door plug on the rear left side of the MAX 9 blew out causing an uncontrolled depressurisation of the aircraft.

The aircraft was able to return to Portland for an emergency landing and of the 171 passengers and six crew on board only three sustained minor injuries, although many more endured a terrifying event that will live on in their memory for a long time.

The MAX 9 has optional rear mid-cabin emergency exit doors that are not used on aircraft with less dense configurations, such as this Alaska Airlines 737-9, where door plugs are installed in their place. Investigations found that the plug had been reinstalled after repairs but not correctly bolted into place and had separated from the aircraft at about 16,000ft.

▲ **The first MAX entered service in January 2016.** Boeing

737 legacy

Boeing's 737 may well take the title as aviation's quietest achiever with billions of passengers riding on the workhorse over more than 50 years. It brought new standards of reliability and comfort when introduced in 1968 and today the 737 MAX has taken that high standard to even loftier levels.

With hindsight, it is easy to say that Boeing should have developed a clean sheet design to follow the 737NG series, but on the other hand it could be praised for using new technology engines on the existing airframe to set many airlines on their way to zero emission flying.

In a similar way to the Comet, the 737 MAX's reputation has been tarnished by the fatal accidents and while it is correct to hold Boeing to account the aircraft has since proved itself to be safe in all operating environments. It should also be remembered that the door plug incident was not a failure of the aircraft but a failure of Boeing's manufacturing oversight.

The aircraft that was derived from a 707 has spawned 13 distinct commercial variants, not counting the versions used by the military, business jet operators and freight carriers. For its innovation, reliability and sheer longevity, the Boeing 737 sits close to the top in this discussion of airliner icons.

▼ **Korean Air has five 737 types in its fleet.** Korean Air

Breaking the BARRIER

Of all the icons, Concorde is the one that got hearts racing and cameras clicking as it streaked down the runway.

Aviation is built around technology and all the aircraft featured here have set new standards of excellence that were game-changing in their time and moved the industry to new heights. While Concorde had technology in spades, the sheer mesmerising beauty of watching it take off or land would have been enough for it to be considered an airliner icon.

Concorde had the perfect fit of elegance and power that has not yet been replicated in a commercial aircraft and the way the airframe was able to control that power and float slowly down to land was a breathtaking sight. It was a technological marvel that captured the imagination of millions, but unlike the Boeing 747 not that many people ever got to experience the thrill first-hand.

As with the A380 that would follow years later, Concorde has many critics who point to the massive investments needed to bring it to life and the fact that only 14 entered service as a sign of its failure.

Concorde was never going to be a mass transit airliner, but it was in service for 27 years and many of the technologies developed for Concorde found their way into other commercial programmes. More than 50 years after it entered service, the world is seeing a revival in supersonic passenger transport and may find answers to the issues that prevented Concorde from fulfilling its full potential.

The need for speed has been more visible in aviation than in almost any other human pursuit and while longer-range jets – such as the 707, DC-10 and 747 – brought anywhere on the planet within reach, flights were long and tiring. Just as the Jet Age had slashed travel times, many saw supersonic flights as the logical answer to connecting cities such as London, Sydney, Delhi, New York, Paris and Rio de Janeiro in just a few hours.

The pursuit of supersonic passenger transport began in earnest in the 1950s, spurred on by the advances in high-speed flight from World War Two and the early Jet Age, as well as the evolution of advanced materials, engines and aerodynamics. The supersonic quest had a stuttering post-war start,

▼ **A British Airways Concorde on display in the UK.** Just Jus/ Shutterstock

blossomed in the 1970s and flamed out at the turn of the century.

Beginning the journey

Given the prevailing political and national sentiment of the 1950s, it is amazing that the British and French governments were able to come together to jointly fund, design and develop Concorde. Both nations were working independently on a supersonic airliner, but by the early 1960s it emerged they had settled on similar designs that ultimately led to a treaty between France and the United Kingdom, rather than a commercial contract between companies, being signed on November 29, 1962, to jointly develop Concorde.

The two governments agreed to share resources for the design, development and manufacture of the supersonic aircraft that became known as Concorde after French President Charles de Gaulle used the word to refer to the project in a speech. The agreement committed the two nations to a 50:50 share of Concorde's research and assembly costs.

Construction of two prototypes started in February 1965 with 001 being built by Aérospatiale in Toulouse, France, and 002 by the British Aircraft Corporation (BAC) at Filton in the UK. The French aircraft first flew on March 2, 1969, and the one built in the UK took off shortly after, on April 9, 1969, with both making their public debut at the 1969 Paris Air Show.

In October that year, Concorde 001 broke the sound barrier reaching Mach 2 on its 102nd test flight, with 002 doing the same a week later. To build up interest both set off on international sales and demonstration tours, with 001 performing the first transatlantic crossing and 002 touring the Middle East, Asia and the US.

▲ **Air France Concorde with gear extended.** Herget Josef/Shutterstock

▲ **Rolls-Royce/ Snecma Olympus 593 turbojet.** Tom Meaker/Shutterstock

In many respects, Concorde was a flying laboratory as designers and engineers on both sides of the English Channel continually solved significant challenges on the run and worked out solutions that were way ahead of anything available at that time.

Concorde's powerhouse

Powered by four Rolls-Royce/ Snecma Olympus 593 turbojets, Concorde had a cruise speed of 1,177kts (Mach 2.04) and with an MTOW of 408,010lb it could carry 120 passengers 3,900nm at a ceiling of 60,000ft. Each engine produced 31,000lbf when dry and 38,050lbf with the afterburners in use during take-off and acceleration through the sound barrier.

Hidden under the wings, the four engines do not always get the attention they deserve for the part they played in getting Concorde off the drawing board and flying passengers for nearly 30 years. The Olympus 593 turbojet was initially a joint project between French company Snecma and the UK's Bristol Siddeley Engines Limited (BSEL), which was acquired by Rolls-Royce in 1966 and renamed as the Bristol Engine Division of Rolls-Royce.

The Concorde partners decided that the cost of developing a clean sheet engine would be prohibitive and settled on producing a civilian version of the Olympus 22R, a military engine designed for the BAC TSR-2 strike and reconnaissance aircraft, redesignated as the 591.

The specification for the (now) 593 was finalised in January 1964 with BSEL responsible for development of the engine, Snecma for the exhaust nozzle incorporating the thrust reverser, noise attenuation and reheat, while BAC looked after the variable intake and overall engine installation. The afterburners (reheat) provided a 20% thrust increase so Concorde could take off from existing runways and for transonic acceleration from Mach 0.95 up to Mach 1.7, after which the aircraft could fly supersonically without reheat.

Throughout its development, Concorde continually pushed technology forward and that was certainly the case with the 593's variable-geometry intake system that delivers air to the engine. The turbojets required a precise airflow regime and at Mach 2 intake air had to be slowed from supersonic to

▶ **Avionics and instrumentation in the Concorde cockpit.** Jurav/ Shutterstock

subsonic speeds before entering the compressor.

Concorde's computer-controlled, variable geometry intakes used moving ramps and spill doors to manage shock waves and prevent compressor stalls. The intake system had to react instantly to changes in airflow to prevent engine surges and it relied on a digital signal processor in the Air Intake Control Units to accurately calculate the necessary engine speed scheduling to ensure an adequate surge margin under all conditions.

This technology was later refined in high speed military jets, including the Eurofighter Typhoon and Panavia Tornado, which adopted advanced variable geometry inlets for optimal engine performance across a wide speed range. Concorde also pioneered intake shock wave management, which has recently featured on experimental aircraft such as NASA's X-43 and X-51 hypersonic text platforms.

Supersonic challenges

To operate nonstop transoceanic flights, Concorde needed the greatest supersonic range of any aircraft then available and it achieved that through a combination of powerful engines that were efficient at twice the speed of sound, a slender fuselage with high fineness ratio and a complex wing shape for a high lift-to-drag ratio.

On its flagship Paris to New York route, the flight time was around three-and-a-half hours compared with around seven hours on a subsonic airliner, with Concorde crossing the Atlantic using high altitude dedicated oceanic tracks separated from those used by other aircraft.

▲ **Plenty to look at in a Concorde cockpit.** Antonis Kousoulas/Shutterstock

▼ **It's time for new dreamers to honour Concorde's supersonic legacy.** NARA

▶ **Concorde at Hermeskeil Airport in Germany.** Joseph Creamer/Shutterstock

The aircraft's elegant appearance came from its slender delta wing, ogee platform and droop nose, but these were carefully refined solutions to the aerodynamic challenges of sustained supersonic flight; the wing and platform designed to maintain lift at both subsonic and supersonic speeds without excessive drag, a key feature for Concorde as it switched between those modes of flight.

Design and engineering ideas quickly cross between civil and military aircraft and many of Concorde's design principles informed later electronic control systems for high performance fighter jets, such as Dassault's Rafale. The delta-canard configuration of the Rafale benefits from the lineage with relaxed stability made manageable through digital flight control, a concept to which Concorde's engineers contributed key aerodynamic data.

Pure supersonic fighter jets are optimised for short bursts, whereas Concorde needed efficiency in the long-range cruise at Mach 2. This gave understanding of vortex lift generated by the wing's leading edge that was directly applicable to the Rafale and Eurofighter Typhoon with both fighters exploiting controlled vortices to enhance manoeuvrability.

Sustained supersonic flight causes the aircraft skin to heat up considerably, particularly on the nose with skin temperatures exceeding 120°C, or 248°F. As the fuselage heated up it expanded by as much as 12in, which led to a gap opening up on the flight deck between the flight engineer's console and the bulkhead.

To combat the heat stress, an advanced aluminium alloy was used throughout the aircraft and the majority of the aircraft's surface had to be covered with a highly reflective white paint to avoid overheating the aluminium structure, reducing the skin temperature by 6°C to 11°C (52°F).

▼ **Concorde on display in New York City.** Phil Emmerson/Shutterstock

Concorde pioneered fly-by-wire (FBW) control augmentation in civil aviation and while it was not a fully digital FBW it had mechanical controls assisted by electronic and hydraulic systems to maintain stability at high altitudes in thin air. It was not the same as today's digital FBW but used an analogue architecture to replace conventional mechanical linkages with electrical signalling to hydraulic actuators.

The airliner also used its fuel as a trim system by pumping it between tanks to maintain the correct centre of gravity as speed and altitude

changed. As Concorde accelerated to supersonic cruise, fuel was transferred rearwards to reduce trim drag, and during deceleration it was pumped forward again.

Operating Concorde

British Airways and Air France were the only two airlines to operate Concorde at supersonic speeds and yet at the time of the first flight in 1976 the consortium had received orders, or more correctly non-binding options, for 74 aircraft from 16 airlines. The project had been launched with sales predictions for more than 350 aircraft, but in the end only 14 entered service.

Launch customers Air France, BOAC/British Airways and Pan Am each put their name down for six each, while others to place options included Continental Airlines, TWA, American Airlines, Qantas, Air India, Middle East Airlines, Sabena, Braniff, Eastern Airlines, Lufthansa, Iran Air, Japan Airlines and Air Canada.

After starting service in 1976 with Air France and British Airways, Concorde operated a variety of routes including British Airways services from London to Bahrain, New York, Miami and Barbados and Air France routes from Paris to Rio de Janeiro, Washington, New York and Caracas via the Azores.

US carrier Braniff International Airways operated Concorde at subsonic speed between Dulles International Airport and Dallas Fort Worth International from January 1979 until May 1980. Supersonic flights over land were banned in the US and for regulatory reasons stickers showing a US registration were placed over the French and British registrations of the aircraft for these flights.

These services were essentially Dallas-Washington-London or Paris flights, with the domestic US sectors operated by Braniff flight and cabin crews and Air France or British Airways crews taking over at Dulles International, the US registration stickers removed. Calling it a gimmick is too harsh but the aim of the three-way partnership was to promote Concorde in the US, with Braniff fully reimbursed for any losses it incurred.

Aviation enthusiasts will have seen images of a Singapore Airlines Concorde that briefly operated services between London and Singapore in December 1977 and between January 1979 and November 1980. This was also

▲ **Concorde G-BOAB on take-off climb.** John Selway/ Shutterstock

a marketing promotion – the Singapore Airlines livery on one side of the aircraft and the British Airlines livery on the other.

The service had a technical stop in Bahrain for refuelling and all flights were operated by British Airways pilots and flight engineers. British Airways and Singapore Airlines cabin crew shared services with one team flying the London-Bahrain sector and the other the Bahrain-Singapore sector, alternatively.

The aircraft, operations, flight crew and insurances remained solely under British Airways and the aircraft was not wet leased to Singapore Airlines or flown under its operators certification. As in other parts of the world, noise issues and political factors interrupted supersonic flight and eventually the service ended in November 1980.

Calling time

In 1967, Concorde's development costs were estimated to be around £2bn (£19bn now) and from that 20 were built but only 14 aircraft entered service operated by the flag carriers of the nations that funded the project, the UK and France.

As soon as governments banned supersonic flight over land, the clock started ticking on Concorde's retirement. Much of the supersonic promise depended on long-haul flights from Europe to Australia, Asia, the Americas and the Middle East. Connecting these destinations supersonically became impossible and the business case for Concorde went out the window.

The original airlines that placed purchase options operated long-haul international services and without the advantage of slashing flight times there was little incentive to purchase the low-capacity Concorde. By widening its markets to include more leisure and holiday passengers, British Airways was able to turn a profit with Concorde but with fuel costs and environmental concerns rising the tide was turning.

The fleet's demise was hastened on July 25, 2000, when an Air France Concorde operating flight 4590 from Paris Charles De Gaulle Airport to

▼ **Air France Concorde at Paris Charle de Gaulle Airport.** travelview/ Shutterstock

New York crashed on take-off. All 109 passengers and crew were killed as well as four people on the ground in what was the aircraft's only fatal accident.

After the accident, Air France and British Airways grounded their Concorde fleets and several modifications were made, including lining the fuel tanks with Kevlar material, fitting more secure electrical controls and specially developed burst-resistant tyres. The aircraft returned to service in November 2001, but the September 11 attacks in the US that year had significantly affected transatlantic business travel.

On April 10, 2003, Air France and British Airways simultaneously announced they would retire Concorde later in the year, citing increasing maintenance costs and falling demand. Air France operated its final commercial flight from Paris to New York on May 30, 2003, while British Airways retired its fleet in October 2003.

▲ **There are 16 Concordes on public display.** Colinmthompson/Shutterstock

The legacy

The dream of flying faster than the speed of sound lives on and, in 2026, US company Boom Supersonic is steadily developing its Overture aircraft for its first flight in the early 2030s. Boom faces the same issue that bedevilled Concorde – the ban on supersonic flights over land – and although Boom has developed a promising solution, authorities have not yet set allowable noise limits.

There has never been another aircraft that can evoke such attention and excitement as Concorde and more than 20 years since it left the skies its influence and legacy lives on in today's civil and military aircraft. It set the model of cross-country collaboration that Airbus was founded on and spawned many new technologies that were specifically developed to meet the supersonic challenge.

As a product of the 1960s, its environmental footprint would not meet today's standards and the exclusivity around who flew on it is unlikely to be repeated. Despite all that, it was an outstanding technical success and this iconic aircraft is a testament to the aviation spirit shown by pioneers such as the Wright Brothers more than a century ago.

The Jumbo Era

In 1975, a Qantas 747 introduced me to international travel with a five-stop flight to Frankfurt and the magic of the jumbo has never left me.

▶ **Pratt & Whitney 747 engine with an unlikely visitor.** Boeing

This collection of airliners aims to catalogue those that have shaped commercial aviation over the last 100 years and in that regard the Boeing 747 has a very strong case to be ranked number one as the most iconic of all.

From a technical, production, passenger experience and marketing aspect it completely changed the game and instantly brought air travel within the reach of the masses, opening opportunities and markets globally. The foresight of the people at Boeing was amazing and it was their willingness to involve airlines, like Pan Am, in the design process that inspired the 747's ultimate success.

Jumbo beginnings

Somewhat ironically, the genesis of the 747 was not the very successful 707 that preceded it but rather a 1963 United States Air Force (USAF) project for a very large strategic transport aircraft with a load capacity of 180,000lb and a range of

▼ **An exciting moment at San Francisco International in 2013.** Bill Larkins CC BY-SA 2.0 via Wikimedia Commons

▲ **Qantas was the only 747-400ER operator.** Qantas

5,000nm with a payload of 115,000lb at a cruise speed of 430kts.

Boeing, Douglas and Lockheed were the frontrunners and to meet the requirement for a front cargo loading door all placed the cockpit above the cargo area, giving birth to the distinctive hump shape of the 747. The contract was awarded to Lockheed for the C-5 Galaxy, but the idea of a door at the front and a cockpit above was soon to appear on the 747 drawing boards at Boeing.

Fuelled by the success of the 707 and Douglas DC-8, jet travel – with its much shortened flight times, smoother rides and cheaper tickets – had sparked a boom in passenger demand. Juan Trippe, the founder of Pan Am, had introduced the 707 and DC-8 to the airline but with demand surging he approached Boeing to build a much larger jet capable of operating international routes and reducing seat-mile costs to open jet travel to the masses.

Boeing took up the challenge and assigned the project to engineer Joe Sutter, who headed the 747 design and build team from 1965 until the aircraft roll-out in 1969, earning him recognition as the 'Father of the

▲ **Up close to a KLM 747 in Sint Maarten.** Aperissinotti CC BY-SA 4.0 via Wikimedia Commons

▼ **British Airways 747 adorned in the BOAC livery.** British Airways

747'. Sutter consulted heavily with Pan Am and other airlines to gain more understanding of their needs at a time when many believed supersonic flight would ultimately dominate future air travel.

Hedging its bets, Boeing designed the 747 with that in mind and built it to be easily converted to a freighter if passenger numbers did not meet expectations. In April 1966, Pan Am stepped up with an order for 25 747-100s, which gained the airline significant influence over the design and development of the aircraft, something never seen previously. With the first to be delivered by the end of 1969, Boeing was always in a race against time, particularly as it did not have a facility large enough to build the 747, but somehow it made that tight deadline.

To house 747 production, Boeing decided to build the world's largest building (by volume) on the north side of Paine Field about 22 miles from Seattle in Washington state. Construction began in 1966 and the facility was officially opened in May 1967; by September 1968, the first 747 rolled off the production line.

Time to fly

The first 747 operated its maiden flight on February 9, 1969, and while it performed well the programme was delayed by issues with the Pratt & Whitney JT9D engines – including engine stalls caused by rapid throttle movements and distortion of the turbine casings after a short period of service.

To show off its new and stunning jumbo jet, Boeing unveiled it to the public at the 1969 Paris Air Show and in December that year the FAA granted Boeing the 747's certificate of airworthiness. Between 1967 and 2023, more than 100 customers purchased 1,574 aircraft that logged more than 118 million flight hours and nearly 23 million flight cycles.

Boeing did exceptionally well to meet its promise by delivering Pan

Am's first 747 on December 12, 1969, and on January 15, 1970, it was officially christened by First Lady Pat Nixon at Dulles International Airport. The 747 era was about to start and on January 22, 1970, Clipper Victor entered service on Pan Am's flagship New York–London route, substituting for Clipper America, which had been scheduled for the launch but was unavailable.

When the 747 entered service in 1970, its capability to carry more than 400 passengers on long-haul flights, compared with around 180 on the 707, made it the largest-capacity airliner of its time. With a twin-aisle cabin, the 747 was the world's first widebody and its early operators included Pan Am, Trans World Airlines, Japan Airlines, Lufthansa, BOAC, Air France and Northwest Orient.

▲ **Flight engineer's 747 station on a 747-200.** Felix Stember CC BY-SA 3.0 via Wikimedia Commons

The variants

Over its seven decades of production, the 747 spawned five main passenger variants, the 747-100, -200, -300, -400 and -8I, with multiple sub-types such as the 747-SP and the 747-200M combi that had a removable partition on the main deck separating the passenger and cargo sections.

The 747's shape remained largely unchanged across the types although successive variants delivered meaningful gains in range, payload, efficiency and crew requirements. The changes kept the 747 current for decades and were designed to match the shifts in demand, network planning and airline economics between 1970 and 2023.

Across all variants, the 747 had typical ranges between 4,620nm on the -100 to 7,730nm on the -8I, with each variant generally offering increased range, especially on the -400 and -8I types. The shorter SP was designed for long-haul routes with fewer passengers and had a range of 6,710nm.

The 747-100 was the original variant to enter service in 1970 and was soon followed by the -200 in 1971 and the 747SP in 1973. The first 747 was powered by four Pratt & Whitney PW JT9D-3A engines, had an MTOW of 735,000lb and originally an upper deck with just three windows either side to accommodate a lounge area.

As airlines saw the opportunity to add a premium experience in the upper deck, Boeing offered an upper deck with ten widows on either side as an option and some of the early -100s were retrofitted with the new configuration. Within the -100 series, Boeing added two sub-types, the 747SR short-range version for Japan Air Lines and the 747-100B with increased range and higher MTOW.

▼ **Thai Airways 747-400 at Denpasar Airport in Bali.** Bahnfrend CC BY-SA 4.0 via Wikimedia Commons

▶ Flying in a 747 was a whole new world in the 70s.
Qantas

▶ After a long flight from Australia it's time for a tidy-up.
Qantas

Boeing delivered 205 of the -100 series, including nine 747-100Bs and 29 747-100SRs. Once airlines saw the possibilities of the 747 they wanted to extend route lengths so Boeing introduced the -200, which increased fuel capacity and offered more powerful engine choices, including the PW JT9D-7, GE CF6-50 and Rolls-Royce RB211-524B powerplants.

The increased power and fuel capacity lifted the -200's range to 6,500nm with the MTOW increased to 833,000lb. Depending on the layout, passenger seating increased by ten seats with the maximum capacity certified to 490 seats.

The 747-200M (originally designated -200BC) could carry up to 238 passengers in a three-class configuration with cargo carried on the main deck. A total of 393 747-200 versions were built before production ceased in 1991. The majority of those were 747-200B with 225 delivered, followed by 78 of the -200M, 73 -200F freighters and 13 -200C combi aircraft. Four 747s were delivered to the military.

The 747-300 series made more incremental improvements, but the big boost to 747 sales came with the longer-range 747-400, which Boeing announced at the 1984 Farnborough Airshow. The -400 made its first flight in April 1988, received type certification in January 1989 and entered service with launch customer Northwest Airlines service in February 1989.

Airlines such as Cathay Pacific, KLM, Lufthansa, Singapore Airlines and British Airways announced orders soon after its 1985 launch, followed by United Airlines, Air France, Malaysia Airlines and Japan Airlines. The 747-400 would go on

▼ Qantas used the 747SP on nonstop transpacific flights.
Qantas

to become the definitive 747 type, accounting for 44% of all jumbo jet sales with 1,573 delivered.

The -400s wingspan was stretched by 17ft through wingtip extensions and the horizontal tail was redesigned to fit a 3,300 US gallon fuel tank with new, more powerful and fuel efficient engine options from Pratt & Whitney (PW4000), General Electric (CF6-80C2) and Rolls-Royce (RB211-524G/H).

The longer-range 747-400 typically accommodated 416 passengers in a three-class layout over a range of 8,383nm with an MTOW of 875,000lb. The type was offered in passenger (-400), freighter (-400F), combi (-400M), domestic (-400D), extended range passenger (-400ER) and extended range freighter (-400ERF) versions.

The 747-400D was designed for short-range operations with maximum seating for 624 passengers, which was achieved by removing the upper deck galley. Boeing delivered just 19 of these variants to Japanese carriers All Nippon Airways and Japan Airlines.

The 747-400ER was an interesting example as it attracted just six orders from one customer, Australia's flag carrier Qantas. The -400ER was launched in November 2000 after Qantas ordered six and with an MTOW of 910,000lb the aircraft featured structural modifications and modified landing gear.

The -400ER had an option for one or two additional 3,240 US gallon body fuel tanks in the forward cargo hold, but Qantas opted for the single tank configuration. The changes added 435nm range or 15,000lb more payload compared with the 747-400 aircraft. The -400ER was specifically ordered by Qantas to operate full loads between Melbourne in Australia and Los Angeles, particularly on the westward flights through unfavourable winds.

▲ **Cabin crew from the airlines that initially ordered 747s.** Boeing

▼ **Japan Airlines 747 taking off from Zagreb in 2008.** Mulag CC BY-SA 3.0 via Wikimedia Commons

Farewell flight engineer

Perhaps the most far-reaching aspect of the 747-400 occurred on the flight deck with a new cockpit obviating a need for the flight engineer role, something Boeing had already done on its 757 and 767 aircraft. Rather than just cutting out the flight engineer (FE) Boeing redesigned the cockpit using deep automation to monitor and diagnose vital systems to reduce the pilot's workload.

The 747-400 was designed and certified for a two-pilot crew with the addition of an advanced glass cockpit, integrated avionics and automated systems eliminating the role of the FE. The two-person flight crew presented airlines with significant cost savings and positioned the 747-400 more closely with the Airbus A300, able to remain competitive with increasingly efficient twin-engined airliners.

Earlier 747 versions featured a dedicated FE station behind the pilots crammed with gauges and dials that the FE continuously monitored, scanning the state of engines, fuel, electrics, hydraulics, pneumatics and pressurisation. The FE's primary role was to spot subtle irregularities before they became failures, but on the -400 those panels and the dedicated station were completely removed, cutting the cockpit's number of gauges, dials and knobs from 971 to 365. The state-of-the-art glass cockpit was built around six large cathode ray tube (CRT) displays that consolidated the information previously spread across all those analogue panels.

The most significant change was the Engine Indicating and Crew Alerting System (EICAS) that monitored systems and automatically alerted the crew when something went wrong. The EICAS prioritised warnings, cautions and advisories by their severity, which simplified the crew's workload by eliminating the need to interpret multiple raw parameters during abnormal situations.

The EICAS also displayed system synoptic pages that allowed pilots to instantly see which components were affected and how they were configured. In simple terms, EICAS replaced the FE's monitoring and diagnostic role with real-time automated logic. The 747's legacy is heightened by demonstrating that well-designed automation could safely replace a role that had been fundamental to commercial aviation.

One last try

Apart from Concorde, most of these iconic airliners ran out of time when new technology was developed or market demands changed, such as the shift to hub and spoke and then back to point-to-point networks. The economics of operating older and less fuel-efficient, quad-engined aircraft ruled time on the 747 when new twin-engined widebodies like the 777 and A330 gained acceptance.

In 2005, Boeing decided to inject new life into the 747 by re-engineering it into the 747-8, although the OEM took a safety-first approach by initially launching the type as the 747-8F freighter. The 747 platform had been adopted by many cargo airlines and logistics companies that had developed extensive infrastructure and operational knowledge around the aircraft.

With an additional 16% payload capacity and an MTOW of 987,000lb, the -8F was a winner from the start, although as 747s already carried around 50% of global air freight it started from a very strong base. The 747-8F was launched in 2005, had its first flight in February 2010 and was delivered to Cargolux in October 2011.

The concept behind the 747-8 was to marry the proven 747 platform with the new-generation engines and systems of the 787 Dreamliner, producing a quieter, more economical and more environmentally friendly aircraft. It is the first stretched version of the 747 and with a length of 250ft 2in compared with the standard 747 length of 231ft 10in.

The passenger version, named the 747-8 Intercontinental or -8I, was designed to carry up to 467 passengers in a typical three-class layout with a range of 7,370nm. It was powered by new GEnx-2B engines that reduced fuel usage, noise and emissions with improved performance over earlier 747 types.

The 747-8I was given a 330-minute ETOPS rating from the FAA, which was the first time a quad-engined aircraft received this design approval, enabling operators to fly on virtually any city pair

▼ **Cathay Pacific operated both passenger and freighter 747s.** Cathay Pacific

routing. It had upgraded avionics and flight management systems and new wings with the cabin heavily modelled on the 787 Dreamliner.

The first 747-8I flew in March 2011 and was certified by the FAA in December 2011 with Lufthansa starting the first services on June 1, 2012. The aircraft entered service when airlines were moving away from large, four-engined aircraft and it proved to be more popular with cargo operators such as Cargolux, Cathay Pacific Cargo, UPS, Atlas Air and Silk Way West Airlines.

Airlines were less enthusiastic with only Lufthansa, Korean Air and Air China ordering the 747-8I, although it found some success in the VIP and private aircraft market, including with the governments of Brunei, Kuwait, Saudi Arabia, United Arab Emirates and Qatar.

With the market disappearing in 2020 due the impact of the pandemic and declining demand, Boeing announced the end of 747 production and the final 747-8F was delivered to Atlas Air in January 2023. In total, there were 155 of the 747-8s delivered, of which 107 were for freighters and just 48 for passenger services. It is still common to see 747F aircraft at global hubs in Asia, Europe, the Middle East and North America and that they will be around for many more years is a testament to the longevity and durability of the 747.

The 747 transformed commercial aviation by making international travel accessible and affordable to the masses and by introducing the twin-aisle template for widebody airliners that still dominates aviation today. It was key to developing hub and spoke passenger networks and the backbone of air freight operations for decades and the lead for new types like the 777, A350, 787 to be developed.

This airliner icon has lasted for more than 50 years and few who saw it will ever forget the wonder of watching it take off or land, especially for the first time in the early 1970s.

▲ **The 747-8I entered service with Lufthansa in 2012.** Lufthansa

▼ **British Airways 747s drew a lot of attention at Heathrow Airport.** British Airways

Time for THREE ENGINES

OEMs generally listen when customers want a new airliner because they see the opportunity to steal a march on their competitors, as happened with the two widebody trijets.

In a similar vein to how Pan Am founder Juan Trippe saw the need for the Boeing 747, it was a direct approach from American Airlines in the late 1960s that spurred the introduction of the widebody trijet era.

American Airlines (American) wanted an airliner that could carry 250 passengers on transcontinental routes and invited US aerospace companies Lockheed and Douglas to submit proposals. The two organisations accepted the challenge that resulted in the birth of the McDonnell Douglas DC-10 and the Lockheed L-1011 TriStar, with both aircraft operating their first flights in August and November 1970, respectively.

▶ Air New Zealand DC-10 departing Sydney, 1977. Daniel Tanner CC BY-SA 4.0 via Wikimedia Commons

▲ **Orbis Flying Eye Hospital DC-10.** Orbis

◀ **Delta Air Lines Lockheed L-1011.** Aero Icarus Zurich CC BY-SA 2.0 via Creative Commons

The trijets bridged a gap in the market between early twin-engined aircraft and the quadjet 747 and allowed airlines to profitably operate high-density and long-haul routes at a time when regulations and engine reliability constrained long overwater flights for twin-engined aircraft.

Somewhat ironically, it was advances in engine technology and reliability that led to changes in the safety standards for twin-engined aircraft, foreshadowing the end of the trijet passenger era, although many lived on for years to come.

ETOPS opportunity

In 1936, the US Bureau of Air Commerce, the predecessor to today's US Federal Aviation Administration (FAA), limited commercial operations to within 100 miles of an 'adequate' airport, which for many in-service aircraft equated to around one hour of flying with one engine inoperative.

The International Civil Aviation Organisation (ICAO) sets the standards for Extended-range Twin-engine Operations Performance Standards (ETOPS), which are designed to ensure that a twin-engined airliner can safely reach

▲ **Thai Sky Airlines L-1011in Bangkok, 2006.** Mark Tang GFDL 1.2 via Wikimedia Commons

a diversion airport with only one engine operating.

The evolution of jet technology that could reliably produce more power than piston engines, such as the Pratt & Whitney engines used on the McDonnel Douglas DC-9 and Boeing 727, meant that by the late 1960s most large civil airliners were jet powered. These advancements resulted in a 1964 rule change that waived the 60-minute rule for three-engine aircraft, opening the opportunity for widebody trijets to enter service.

This left only twin-engine jets restricted by the 60-minute rule, enabling trijets and quadjets to dominate international long-haul flights well into the 1980s. Engine and aircraft technology kept advancing and in 1985 the FAA increased ETOPS limits to 120 minutes, opening the door for twin-engined airliners, such as the 767 and A300, to build a presence in the market at the expense of the trijets, particularly on routes where the four-engined 747 was not a viable commercial option.

Lockheed's return

In 1958, Lockheed delivered its first Electra L-188 four-engine turboprop, but with only 170 built, production ended when the final aircraft was delivered to Garuda in 1961. The L-188 was involved in numerous fatal accidents and other safety incidents and it ended Lockheed's civil programme until the L-1011 was introduced in the 1970s.

Lockheed focused on its military programmes and had notable successes with the C-130 Hercules and SR-71 Blackbird, but when the approach came from American Airlines it jumped at the chance to re-enter the civilian market in direct competition with McDonnell Douglas.

With such a distinguished military pedigree, it is unsurprising that Lockheed set out to create something special and challenged its designers and engineers to develop cutting-edge technology that set new standards for commercial airliners.

Conversely, the project was a financial disaster for Lockheed that was exacerbated by the failures of Rolls-Royce, which teetered on the edge of bankruptcy as the programme fell nearly a year behind schedule.

The first TriStar entered service in April 1972 with US carrier Eastern Air Lines, but with only 250 delivered, around half of what was needed to break even on the programme, production ceased in 1984. Some of the operators included Delta, TWA, Pacific Southwest Airlines, Air Canada, Cathay Pacific, Aer Lingus, Air Lanka, Royal Jordanian, Gulf Air, British Airways, Air Transat and All Nippon Airways.

The most common variants were the L-1011-1/-100 family and the longer-range -500, which was popular among international operators including Delta and British Airways. The -500 was powered by three Rolls-Royce RB211-524 engines with a range up to 5,200nm, a service ceiling of 43,000ft, a cruise speed of around 525kts and a wingspan of 164ft.

▼ **Lockheed L-1011 TriStar Stargazer, 2019.** Ben Smegelsky Public Domain via Wikimedia Commons

▲ **L-1011 flown by Faucett Peru Airlines.** Aero Icarus Zurich CC BY-SA 2.0 via Creative Commons

The most visible difference to the DC-10 is the way the rear engine was positioned within the fuselage using an S-duct configuration to minimise drag, ensure maximum aerodynamic efficiency and reduce noise. Lockheed selected the Rolls-Royce RB211 as the only engine option, but when Rolls-Royce went in to receivership in 1971 it was left with nowhere to turn.

The L-1011 design had been frozen and with no comparable engine available to suit the airframe Lockheed had no other option but to endure the significant delays while the UK and US governments bailed out the programme with subsidies and guarantees. However, this 12 month delay gave the DC-10 a significant head start that proved to be an insurmountable hurdle for the TriStar to overcome.

The TriStar operated with a partial fly-by-wire system and had a highly advanced autopilot with a unique direct lift control (DLC) that produces smooth landing approaches without relying on significant pitch changes while on the approach path. It was the first widebody to achieve FAA certification for Cat IIIB autolanding, which approved the TriStar for completely blind landings performed by the autopilot in zero visibility weather.

The innovative Avionic Flight Control System (AFCS) included speed control, a moving map navigation system and DLC with all the systems integrated to allow the aircraft to complete a full flight with minimal pilot inputs. This was a completely new capability and to prove the point a test flight was performed on May 25, 1972, from Palmdale, California, to Dulles Airport outside Washington DC.

The TriStar, flown by test pilots Anthony LeVier and Charles Hall, carried 115 crew, employees and reporters on the four-hour flight, with the AFCS engaged from take-off roll to landing. Lockheed described it as groundbreaking and the first cross-country flight without the need for human hands on the controls, adding that "fly-by-wire technology was here to stay".

Plagued by safety issues

When American Airlines requested proposals from Lockheed and McDonnell Douglas for a new type of aircraft, the prevailing wisdom was the L-100 TriStar had a head start. The TriStar offered many groundbreaking technologies, new levels of passenger comfort and reduced costs with the new Rolls-Royce RB211 engine.

Any confidence that Lockheed had was knocked out in February 1968 when American Airlines announced its intention to order the McDonnell Douglas DC-10 and power it with the Rolls-Royce RB211 engine. In a further blow to Lockheed, United Airlines joined in by ordering 30 DC-10s with options to purchase a further 30 aircraft.

These early DC-10 successes set off vigorous competition between the two trijets with price often used as the significant difference between the two brands, leading to poor financial returns for Lockheed and McDonnell Douglas, with the DC-10 outselling the L-1011 by nearly two to one.

The DC-10 also had a one-year head start on the L-1011 due to the Rolls-Royce troubles and from there Lockheed was always playing catch-up to its rival. McDonnell Douglas avoided all those issues by ultimately selecting the GE CF6-6 engine, the first civil variant from the CF6 product line.

▼ **TWA Lockheed L-1011 at London Heathrow, 1987.** Airlinephoto/ Shutterstock

▶ **Continental 1973-era DC-10 cabin.** clipperarctic CC BY-SA 2.0 via Creative Commons

▶ **Japan Airlines DC-10 JA8542 in 2003.** Ken Fielding CC BY-Sa 3.0 via Wikimedia Commons

The first DC-10 entered service with American Airlines on August 5, 1971 and the 446th and final aircraft was delivered to Nigeria Airways in July 1989. McDonnell Douglas was not done with trijets and production moved on to the MD-11, an extended version of the DC-10 that was grounded in 2025 due to safety issues.

Over its lifetime, multiple variants of the DC-10 were developed with the longer-range DC-10-30 being the most popular capturing close to 200 orders. This variant had a typical load range of around 5,400nm due to its larger fuel tanks and wingspan and the GE CF6-50 engines.

A DC-10-30ER was also produced that pushed the range out to around 5,700nm with an extra fuel tank in the rear cargo hold. Finnair took delivery of the first ER aircraft in 1981 and it is interesting to note that in 2026 Qantas will take delivery of a bespoke Airbus A350, also fitted with extra fuel tanks, to fly nonstop from Sydney to London and New York.

The DC-10-30CF also spawned a convertible cargo/passenger version, along the lines of the Boeing 747SP as well as an all-freight DC-10-30AF aircraft, but neither of these gained popularity. However, the DC-10-30AF was a success when adopted by the US Air Force in the 1980s as the KC-10 Extender, a military version used for aerial refuelling and more recently operating as aerial firefighting tanker and as a flying Eye Hospital.

Unfortunately, no discussion of the McDonnell Douglas DC-10 can avoid talking about its chequered safety record, particularly the three catastrophic crashes in the 1970s of aircraft operated by Turkish Airlines and American Airlines.

On June 12, 1972, an American Airlines DC-10 (N103AA) was extensively damaged when the rear cargo door separated from the aircraft at 11,750ft near Windsor, Canada. The US National Transport Safety Board (NTSB) report said the loss of the door caused rapid decompression, which, in turn, caused failure of the cabin floor, and the separated door substantially damaged the leading edge and upper surface of the left horizontal stabiliser. The pilot was able to gain control of the aircraft and return to Detroit to make an emergency landing with only minor injuries to two flight attendants and nine passengers. The NTSB said the probable cause was "improper engagement of the latching mechanism for the aft bulk cargo compartment door". On July 6 that year, and despite this finding, the NTSB only made safety recommendations to modify the door locking system to the FAA, which replied on the following day

▼ **DC-10/MD-10s found roles as freighters.** Tomas del Coro CC BY-SA 2.0 via Wikimedia Commons

that these were "being considered", but "may not be feasible".

On March 3, 1974, a Turkish Airlines DC-10 (flight TK981) departed Paris Orly bound for London and ten minutes into the flight lost its rear cargo door at 11,000ft. The sudden depressurisation and destruction of the floor structure impaired the flight controls to such an extent that the aircraft was uncontrollable. All 346 people on board were killed, including six passengers who were ejected before the crash when the cabin section was torn apart mid-air. The FAA report said the underlying factor causing the accident was "the incorrect engagement of the door latching mechanism before take-off, made possible by the characteristics of its design".

In 1979, an American Airlines DC-10 operating AA191 crashed immediately after taking off from Chicago O'Hare Airport when the left engine and strut assembly separated from the wing, severing hydraulic lines and the subsequent retraction of the left wing outboard leading edge slats. All 271 passengers and crew, as well as two people on the ground, died in the accident which was later found to have been caused by incorrect maintenance procedures.

In an incident the NTSB described as similar to the AA191 crash, on November 4, 2025 an MD-11F operated by United Parcel Service crashed after taking off from Louisville International Airport, killing the three crew and 11 people on the ground. The NTSB preliminary report said surveillance video showed the left engine and pylon separating from the wing shortly after rotation with a fire igniting on the left wing.

On November 8, 2025, the FAA issued an Emergency Airworthiness Directive that prohibited further flight of MD-11 and MD-11F airplanes until they were inspected and all applicable corrective actions were performed using a method approved by the FAA. That was quickly superseded with a new directive that included both MD-11 and DC-10 series airplanes, the latter based on their similar design to the MD-11.

The widebody trijets filled a gap in the market until that gap was plugged by more fuel-efficient twin-engined aircraft that could meet ETOPS certification rules. The fact that McDonnell Douglas and Lockheed produced two quite similar aircraft meant that price became a major determinant for buyers and profits dwindled for both OEMs.

These are airliner icons because they were products of their time, one that offered new solutions and more comfortable experiences for medium and long-haul passengers. The crashes of the DC-10s were unsettling and cast a pall over the aircraft, but as with most of the aircraft in this bookazine it was engine technology that spelt their end.

▲ **Ariana Afghan Airlines near London Heathrow, 1980.** Steve Fitzgerald GFDL 1.2 Wikimedia Commons

Soviets Try Trijets

The Tupolev Tu-154 and Yak-40 had a significant impact on Eastern Bloc aviation by opening up remote regions to reliable air travel.

▲ **Estonian Aviation Museum presents a Yak-40.** Ivo Kruusamagi CC BY-SA 4.0 via Wikimedia Commons

▶ **Tupolev Tu-154 RA-85826 landing in 2011.** Artem Katranzhi Russia CC BY-SA 2.0 via Wikimedia Commons

In the post-World War Two era, the Soviet Union and the Western Alliance were jockeying for aviation leadership as new engine and airframe technologies ushered in the commercial jet age. While both were developing new aircraft for their own captive markets, technological competition, fuelled by the development of supersonic passenger aircraft, became a marker for aerospace leadership and national pride.

The US-produced McDonnell Douglas DC-10 and the Lockheed L1011 TriStar both made their first flights in 1970 while two Soviet trijets, the Yakovlev Yak-40 and Tupolev Tu-154, first took to the air in 1966 and 1968, respectively. The four aircraft entered commercial service between 1968 and 1972 with the Soviet Union manufacturing a combined 2,037 trijets compared with 696 widebody trijets produced in the US.

Flying 'cigarette butt'

Airliners like Concorde and the 747 are admired for their aesthetics and graceful shapes, but the Soviet Yakovlev Yak-40, nicknamed the 'cigarette butt' because of its trademark trail of engine smoke, was more about function than form.

However, it ranks as iconic because it was the world's first jet aircraft specifically designed for regional airline service. Prior to its entry into service, regional air travel in the Soviet Union was largely the domain of turboprop piston-engined aircraft from Tupolev, Ilyushin and Antonov.

Many of these services were to remote regions with unpaved runways so when Aeroflot decided to switch to jet aircraft a key requirement for a new aircraft was ruggedness, durability and the ability to land on short unpaved runways, often in extremely challenging environments.

To meet those demands, the Yak-40 had a single large wheel on each main undercarriage leg and a single wheel for the nose gear, reinforced with high-strength shock absorbers. It was powered by three Ivchenko AI-25 turbofan engines with two mounted on pylons at the rear of the cabin and a third in the rear fuselage.

The jet was designed to be self-sufficient when on the ground at remote airports spanning Siberia and Central Asia, where minimal ground support was available. It had an auxiliary power unit for engine start-up, a set of ventral airstairs in the rear fuselage for passenger boarding and a three-engine configuration that was designed to reduce foreign object damage from unpaved runways.

With a cruising speed of 300kts and a service ceiling of 26,247ft, it was not designed for speed but was optimised for short sector, low-altitude performance on routes where reliability and operational flexibility were paramount. The aircraft was initially designed to be operated by a captain and first officer, but the cockpit was adapted to accommodate a flight engineer/mechanic while a flight attendant looked after the cabin.

The Yak-40 carried 24 to 32 passengers with a maximum range of 1,100 miles, but at maximum payload the range was just 540m. Passengers boarded via the inbuilt airstairs that led to a rear bulkhead door and into a small lobby area that housed the toilet and had space for baggage, which passengers carried on board.

The cabin was pressurised and climate conditioned and while it was relatively basic by Western standards it was a significant step up from what passengers had previously experienced in ageing turboprops. A typical configuration was eight or nine rows of 1-2 seating with a high density configuration of eight rows of 2-2 seats.

The Yak-40 had a forward left door for service access or boarding via an airbridge where available, and emergency exits over each wing and in front of the wing on the right side of the aircraft. The placement of those varied depending on the variant and number of windows on the fuselage, which shifted from eight to nine and up to 11 on some prototype Yak-40s.

The aircraft was in production from 1967 to 1981; more than 1,000 were built and it was the first Soviet-built airliner designed to meet Western airworthiness standards. It formed the backbone of Aeroflot's local operations, serving more than 276 domestic destinations in 1980.

The Yak-40 had no obvious competitor as well optimised for regional flying to remote and under-developed airfields and the Soviets mounted a solid export effort, demonstrating the aircraft in 75 countries, predominantly in Asia, Africa and Latin America.

The Yak-40 was operated in around 35 countries, including Afghanistan, Angola, Albania, Bolivia, Bulgaria, Cuba, Czechoslovakia, Egypt, Kazakhstan, Lithuania, Peru, Libya,

▲ **Severstal Yak-40 RA-88296.** Maarten Visser Nederland CC BY-SA 2.0 via Wikimedia Commons

▼ **Yakolev Yak-40 operated by Poland Airforce.** Bjorn Strey CC BY-SA 2.0 via Wikimedia Commons

▲ **Vologda Air Yak-40.** Sergey Riabsev GFDL 1.2 via Wikimedia Commons

▶ **The Czech Air Force operated a Yak-40.** Arpingstone Public Domain via Wikimedia Commons

Ukraine, Uzbekistan, Vietnam, Venezuela, Russia, Guatemala, Greece, Italy and West Germany.

The Yak-40 may have earned a reputation as the world's slowest jet but it was a product of its time, built for a purpose for which it was ideally suited. It pioneered the concept of a specialised regional jet and paved the way for the many Bombardier CRJ and Embraer ERJ jets that revolutionised regional services globally.

Soviet success story

The Tupolev Tu-154 was born in the same era as the Yak-40 and had a similar mission to serve remote and under-developed airports, as well as operate domestic routes between major cities and regions. It was conceived in the late 1960s and came to life through a decree from the Soviet Union's Council of Ministers in August 1965.

The project called for a new-generation airliner powered by three Kuznetsov NK-8-2 turbofans capable of carrying up to 160 passengers and 18 tonnes of payload over 1,890nm. The initial Tu-154 prototype performed its first flight in October 1968 and entered trial commercial flight, carrying mail and freight only, in 1971 with Aeroflot passenger flights taking off in February 1972.

The Tu-154 was designed as a fast, medium to long-haul, modern passenger jet that could endure the Soviet Union's extreme climates from the frigid cold of Siberia to the intense heat of Central Asia. Capacity changed depending on the variant, but a typical one-class configuration carried up to 180 passengers in a 3-3 layout, while the two-class layout could seat 144 passengers.

Concurrent with the Tu-154, Tupolev had its hands full developing the high profile supersonic Tu-144, which strained the funds and technical resources available for the Tu-154 project. Extensive testing showed a number of design shortcomings but rather than delay the programme further, Soviet authorities launched the aircraft into production while ironing out issues along the way.

For example, the slender and flexible wings of early versions used an advanced aluminium alloy that proved to be prone to frequent cracks and shortened life, and it required structural reinforcements in later versions. The Tu-154B featured a redesigned wing with new aluminium alloys to overcome those fatigue issues.

▼ **Azerbaijan Airlines Tupolev Tu-154.** Aero Icarus Zurich CC BY-SA 2.0 via Wikimedia Commons

The Tu-154 had a maximum speed of approximately 493kts and a fully loaded range of 1,300nm or 2,100nm with maximum fuel loaded. It was powered with three upgraded Kuznetsov NK-8-2U engines that gave the Tu-154 the reliability and performance it needed to operate on unprepared runways.

The characteristics of the Tu-154 proved highly demanding of pilots so early variants operated with a flight crew of five, including captain, co-pilot, navigator, flight engineer and radio operator.

In August 1978, a new version was developed that was ultimately known as the TU-154M, with the first example converted from a production Tu-154B-2 and used for flight testing. The first newly produced Tu-154M took to the air in July 1984 and gained certification in

June 1986, with commercial flights beginning in July 1986.

The new type featured improved aerodynamics and was powered by more fuel-efficient Soloviev D-30KU-154 engines that reduced fuel burn consumption by 15-20% on routes up to 1,620nm. The maximum take-off weight was also increased, enabling the Tu-154M to fly up to 2,808nm with 12 tonnes of commercial payload.

The advanced avionics systems and improved flight characteristics allowed flight crews to be reduced to three with an advanced avionics system including a moving map display. The Tu-154M offered several cabin configurations including one-class options for 164, 176 or 180 passengers or a two-class version with eight in first class and 146 in economy.

By the late 1980s, the Tu-154 was operating for around 52% of Aeroflot's total passengers, transporting hundreds of millions of passengers annually. It was widely used by Eastern Bloc airlines in Bulgaria, East Germany, Hungary, Poland, Romania, Syria, Egypt, Cuba, Mongolia, North Korea, Vietnam, Iraq and Czechoslovakia.

The 1991 break-up of the Soviet Union led to the aircraft gaining popularity with the newly established airlines in the ex-Soviet states. The Tu-154 was flown by operators in more than 40 countries and by carriers as diverse as Sichuan Airlines, Cubana, LOT Polish, TAROM, Air Kazakhstan, Libyan Arab, Mahan Air, Air Koryo, EgyptAir, Air Ukraine and Aeroflot.

When production ceased in 2013, a total of 923 aircraft had rolled off the line while the last Tu-154M sold to an airline was delivered to Samara Airlines in 2007 and the final aircraft was handed over to the Russian Ministry of Defence in February 2013.

The demise of the Tu-154 was hastened by noise restrictions introduced by the European Union and increasing fuel costs that made newer, more fuel-efficient aircraft like the A320 and 737 much more economically viable.

With more than 2,000 aircraft produced between them, the Tu-154 and the Yak-40 were major successes in bringing more efficient and reliable services throughout the vast Soviet Union. These iconic airliners opened up routes to under-developed airfields that laboured with ageing turboprops and thoroughly modernised commercial aviation in the Soviet Union and beyond.

▲ **Pulkovo Aviation Enterprise Tu-154.** Aero Icarus Zurich CC BY-SA 2.0 via Wikimedia Commons

▼ **Aeroflot Tu-15 landing at Lisbon Airport.** Pedro Aragao CC BY-SA 3.0 GFDL via Wikimedia Commons

AIRBUS Comes of Age

Perhaps the biggest achievement of the A300 was bringing all the international partners together into a unified and successful Airbus.

New aircraft programmes are a commercial gamble, with designers, engineers and executives developing airliners that will not enter service for many years and then need to remain relevant for decades into the future to generate a financial return.

It is arguable that the groundbreaking A300 carried more weight than most as the very survival of Airbus depended on the new aircraft becoming an international success. With hindsight, it is easy to see how different today's aerospace industry would be if the A300 and Airbus had failed, particularly in the commercial airline sector.

In 1967, aircraft manufacturers in West Germany, France and the United Kingdom signed a memorandum of understanding to collaborate on the development of a new large airliner. The UK withdrew from the project in April 1969, but the French and West German teams pushed on to form Airbus Industrie GIE in December 1970.

In April 1969, the British government announced it would withdraw from the Airbus project but the managing director of Hawker Siddeley, Sir Arnold Alexander Hall, decided the firm would remain in the project to develop and manufacture the wings. Airbus was formed through an agreement between France's Aérospatiale and the antecedents

to Deutsche Aerospace of Germany on a 50/50 share basis.

Spanish company CASA joined in 1971 and in 1979 Britain joined the consortium via British Aerospace, which Hawker Siddeley had merged into, with the shareholdings split 37.9% to both Germany and France, 20% to the UK and 4.2% to Spain.

The A300 was the consortium's launch aircraft and stands alone for its innovation as the world's first twin-engined widebody. It also marks the successful coming together of Europe's aerospace industries. Since then, Airbus has delivered more than 16,000 commercial aircraft and had close to 8,000 aircraft in backlog at the end of 2025.

Despite struggling for sales early, the programme did not fail and from the relatively humble beginnings the A300 went on to penetrate all major regions, including usurping Boeing, Lockheed and McDonnell Douglas in their home market. Between 1971 and 2007, Airbus delivered 561 A300 aircraft and 255 of the shortened A310 variant that competed with Boeing's 767 airliner.

Airbus comes to life

The Airbus A300 was announced as the world's first twin-engined widebody at the 1969 Paris Air Show and the first prototype went into production in September 1969. In a sign of support for Airbus and the A300, Air France signed a letter of intent to purchase six A300s in September 1970, well before the first prototype had been completed.

Airbus split the major work between West Germany, France and the UK, with French firms producing the aircraft's cockpit, control systems and lower-centre portion of the fuselage and German companies producing the forward, rear and upper part of the centre fuselage sections.

▲ **Delta Air Lines A310-221 N801PA, 1992.** Aero Icarus Zurich CC BY-SA 2.0 via Creative Commons

▼ **SATA International Airbus A310 CS-TRG.** Wo st 01/Wikimedia Commons

▲ **Airbus A300B first flight, 1972.** Airbus

Hawker Siddeley was tasked with manufacturing the wing while Spain's CASA produced the horizontal tail. Elements of the wing were manufactured in the Netherlands. This sharing of work with individual assemblies manufactured around Europe and the UK and assembled in France was revolutionary in aviation and became the model for Airbus that remains in place today.

The first A300 was unveiled to the public in September 1972 followed by its maiden flight on October 28, 1972; the second prototype first flew in February 1973. In total, four aircraft spent nearly 1,600 hours in flight testing before certification in Germany and France was achieved in March 1974, followed by certification from the US Federal Aviation Administration (FAA) in May that year.

The A300 was designed to compete directly with the McDonnell Douglas DC-10 and Lockheed Martin L-1011 TriStar aircraft through being lighter, smaller and more economical to operate than these three-engined competitors. Airbus marketed the A300 extensively using the testing period to undertake a six-week tour of North and South America in September 1973, catching the eye of former astronaut and Eastern Air Lines CEO Frank Borman.

To make the A300 commercially viable, it was vital for Airbus to penetrate the US market, which was staunchly loyal to its home-grown aerospace manufacturers. Frank Borman liked what he saw, was particularly impressed by the potential fuel savings from the more efficient Airbus airliner and decided to give it a chance.

In 1977, Eastern Air Lines struck a very advantageous deal with Airbus to lease four A300s to prove their worth as replacements for ageing DC-9s and Boeing 727-100s. The A300 performed as promised and delivered 30% fuel savings compared with the trijets, prompting an order for 23 aircraft, the first US customer for the A300.

Airbus also focused on the emerging Asian market, which led to discussions with Korean Air about developing a longer-range A300, later to become the A300B4. In September 1974, Korean Air placed an order for four A300B4s with options for two more, becoming

▶ **Moving large parts through small towns was always a sight.** Airbus

▲ **Nigeria Airways A310-200 in Paris.** Michael Gilliand GFDL 1.2 via Wikimedia Commons

◀ **Airbus A300-600F operated by United Parcel Service.** Airbus

▼ **Lufthansa A300B4 at Stuttgart Airport, 2005.** JetPix GFDL 1.2 via Wikimedia Commons

the first non-European international airline to order an Airbus aircraft.

These successes in North America and Asia raised Airbus's profile and opened the door to many more orders during the late 1970s, including from Pan Am, Indian Airlines, Aerocóndor Colombia, China Airlines, Trans Australia Airlines, Thai Airways, Singapore Airlines, China Eastern and Garuda Indonesia.

On May 24, 1974, the first production A300B2 entered service with an Air France Paris–London service and, while early sales were slow, by 1979 there were 81 A300s in service with 14 airlines, with a backlog to produce firm orders for 133 and options for a further 88. Within a decade, Airbus had secured 28% market share (by dollar value) and was ready to develop its second aircraft, the Airbus A310.

The breakthrough in the US in 1977 came when the FAA changed the restrictions for ETOPS, the rules that governed the way passenger jets could operate over water or remote lands. These restrictions were a major factor in the success of trijets, such as the L-1011 and DC-10, which had carved out a solid market on routes where twin-engined jets could not operate due to ETOPS.

▶ **Mahan Air A310 in Dusseldorf, 2011.** Lasse Fuss CC BY-SA 3.0 via Creative Commons

Once Airbus was able to prove the higher reliability of the A300, the FAA granted certification for the aircraft to fly further from an available airfield than previously allowed for any twin-engined jet. This opened up an extensive market for over-water medium-haul flights and, as the A300 was the first twin-engined airliner certified for ETOPS, its horizons immediately expanded.

A versatile platform

The initial A300B1 spawned five main variants: the A300B2 that was the first to enter service followed by the A300B4, A300-600, A300B10 and the A300-600ST. The A300B10 was later to become the A310 and the MRTT military tanker/transport while the A300-600ST lives on as the Airbus Beluga, the whale-shaped super transporter that was used to ferry A380 assemblies from manufacturers around Europe for assembly in Toulouse.

The most popular was the A300-600 that made its first flight in July 1983, gained certification in March 1984 and entered service with Saudi Arabian Airlines in June 1984. The A300-600R offered longer range through an additional fuel tank in the tail and all A300s built from 1989 were this variant, with the last new build going to Japan Air System in 2002.

The A300-600R had a maximum capacity of 345 passengers or 247 in a typical three-class layout, a range of 4,050nm and a cruising speed of 450kts. Its maximum take-off weight (MTOW) was 378,534lbs and a choice of two engines, the General Electric CF6-80C2 or the Pratt & Whitney PW4158.

Airbus delivered 561 A300s and it reports that in November 2025 there are 210 still in operation with the majority of those being used as freighters, including large fleets at FedEx Express and UPS Airlines. Airlines carrying passengers on their A300s are mostly from Iran – Iran Airtour Airlines, IranAir, Mahan Air, Meraj Air and Qeshm Airlines.

Airbus A310 fills a gap

The extensive marketing programme for the A300 revealed significant demand for a lower-capacity cabin aimed at thinner routes where demand was not enough for the A300 or for airlines that wanted to operate higher-frequency schedules.

▼▶ **Monarch Airlines A300-600 in Manchester, 2011.** Aero Icarus Zurich CC BY-SA 2.0 via Creative Commons

▼ **A300 flown by American Airlines in 1998.** Aero Icarus Zurich CC BY-SA 2.0 via Creative Commons

▲ **Monarch Airlines A300 cabin.** Andrew Thomas CC BY-SA 2.0 via Wikimedia Commons

Airbus also wanted to grow its aircraft portfolio. So, to fill those two aims, it began investigating a smaller derivative of the A300B2, which became known as the A300B10MC (MC = minimum change), later to be the A310. Airbus wanted to mitigate any design costs, so differences to the A300 were kept to a minimum and in 1978 at the Hanover Air Show a model of the proposed A310 was unveiled.

The development of the A310 was heavily influenced by Swissair and Lufthansa, which developed a joint specification in June 1978 and shortly after both carriers placed orders for ten aircraft as the launch customers. In contrast to the A300, the A310 got off to a flying start with orders rolling in before its first flight and additional orders from Air France, Iberia, KLM Royal Dutch Airlines, Martinair, Sabena, Air Afrique and low-cost pioneer Laker Airways.

The A310 came in two versions, the medium-range A310-200 and the longer-range A310-300, while a shorter-range A310-100 did not get off the drawing board due to low demand.

The first A310 prototype flight was in April 1982 and it acquired certification in March the following year, by which time Airbus had secured orders and options for 181 aircraft from 15 airlines. Swissair and Lufthansa introduced the A310-200 to service in April 1983 while the A310-300 entered service in 1986 with Swissair.

The -200 had a range of 3,500nm, a cruising speed of 459kts and an MTOW of 317,466lbs with Airbus increasing the MTOW to 361,558lbs and range to 5,150nm on the -300.

▼ **Thai Airways A300-600 at Bangkok Airport, 2011.** Aero Icarus Zurich CC BY-SA 2.0 via Creative Commons

▲ **The A300 was the original MRTT platform.** Airbus

▼ **The A300 also led to the Beluga Super Transporter.** Airbus

The range of the A310-300 made the aircraft popular on transatlantic routes and in Asia as an entry into widebody operations.

The -300 version went on to be the standard production version with its longer range facilitated by additional centre and horizontal stabiliser (trim-tank) fuel tanks, wingtip fences and a computerised fuel distribution system that allows it to be trimmed in flight, optimising the centre of gravity via the Centre of Gravity Control Computer.

This model also introduced wing fences, or winglets, which reduced vortex drag and thus improved cruise fuel consumption by more than 1.5%. It had a new two-person cockpit with an array of six CRT displays in place of the traditional analogue dials and instrumentation.

Part of the family

Airbus has consistently presented its aircraft models as part of the same family and particularly in Asia it has been successful in pairing aircraft, such as the A330 and A350, to give airlines flexibility in route building. The A310 removed the need for a flight engineer and the same Forward-Facing Crew Cockpit was incorporated into the A300-600, bringing commonality that

◀ **A310 HB-IPG flown by Swissair in 1998.** Simon Butler UK CC BY 2.0 via Creative Commons

enabled pilots to have a dual type rating, a feature that lives on today within the Airbus fleet types.

In simple terms, the A310-300 is a shortened version of the A300 with capacity for 243 passengers in economy or 220 in a typical two-class layout, compared with a maximum of 285 and 247, respectively, on the A300-600R.

The demise of the A310 aligned with the introduction of the new-generation A330 in 1994 and no sales of A310s were recorded in the late 1990s. Airbus reports show that 255 A310s were delivered between 1983 and 1998 when the final aircraft was delivered to Uzbekistan Airways in Central Asia.

The same report shows that at the end of November 2025, there were 48 A310s still in service, with 13 of those in Iran, three in the fleet of Afghan Airlines, three with ULS Airlines Cargo, eight with governments and 21 listed as having undisclosed owners.

In the defence space, the Airbus A310 MRT (Multi-Role Transport) has been operated by multiple air forces as a pure transport aircraft and also, converted to the A310 MRTT (Multi-Role Tanker Transport) configuration, providing air-to-air refuelling.

Legacy

The A300 platform was stretched to make the A330 and A340, shrunk for the A310 and modified into offshoots like the Beluga Super Transporter and the MRTT. The development of the platform introduced many technologies and innovations that were later refined to live on in today's modern airliners.

However, perhaps the true legacy is that the A300 carried the weight of this new multinational business that aimed to break the US stranglehold on manufacturing commercial aircraft. The A300 gave Airbus its start in designing, selling and supporting airliners in very competitive and demanding markets, not to mention managing supply chains across multiple countries and workforces.

In the early days, the future looked bleak for the A300, but the determination and imagination of those Airbus pioneers kept the dream alive and without that we would not be enjoying aircraft such as the A320neo, A330neo, A350 and A380. Travellers all around the world are reaping the benefits of the A300's success and that's a legacy hard to beat.

▼ **Inside the body of the Beluga 'flying whale'.** Airbus

A320 Takes

The Airbus A320 is now the most delivered jetliner in history and its story is a testament to the foresight of the Airbus founders.

▲ **Even in the Antarctic the A319 shines.** Glen Jacobson

In October 2025, Airbus broke a barrier many thought impossible when its A320 Family overtook the Boeing 737 to take the title of the most delivered jetliner in history. The aircraft is so ubiquitous that Airbus says one takes off or lands somewhere in the world every two seconds.

The A320 was launched in March 1984 and by the end of November 2025 it had booked 19,495 orders and delivered 12,375 aircraft across the family of narrowbodies, with a further 7,120 in backlog. The 737 has received orders for 17,108 of which 12,327 were delivered by the end of November 2025, leaving 4,781 in backlog.

Reports from the manufacturers show that in the decade from 2015-2024, Airbus delivered 5,480 A320 Family aircraft compared with Boeing's 737 family deliveries of 3,566 in a clear demonstration of Airbus's leadership in the single-aisle segment.

The record-breaking A320neo was the 64th A320 aircraft delivered to Saudi low-cost carrier (LCC) flynas, which operates an all-Airbus fleet of A320 and A330 jets. The A320 has flourished in the emergence of low-cost airlines where the reliability, lower fuel consumption and continuous evolution have driven the success of LCCs such as AirAsia, easyJet, IndiGo, Jetstar and Wizz Air.

Not content to sit back and bask in the success, Airbus has continually developed new technologies to advance the A320 platform and today we are seeing the results of that in the long-range A321XLRs, which both low-cost and full service carriers are using to develop domestic and international routes once flown by widebodies.

The A320 has been around for nearly 40 years and when it entered service with Air France in 1988 it brought new and truly groundbreaking innovations, such as a full fly-by-wire (FBW) digital control system, computer-based electronic flight instruments, glass cockpits, side-stick pilot controls, advanced composite materials and enhanced passenger comfort.

▶ **A Qantas A321XLR in production at Airbus.** Qantas

World Title

It should be remembered that Airbus was only formed in 1970 and yet its leaders had planned on developing a family of aircraft from its inception, which materialised into the A320, A318, A319 and A321 variants.

It is now assembled at centres in France, Germany, China and the US with many global suppliers producing all the assemblies, parts and equipment that have brought such success. Concurrently, Airbus has brought the A220, A330, A350 and A380 programmes into service and has navigated the rough waters of the COVID-19 pandemic and the supply chain disruptions of recent years.

Based on the numbers, Airbus has overtaken Boeing as the leader of single-aisle commercial airliners and is now in a very sweet spot having developed the 'new engine option' (neo) A320/A321 aircraft that kicked the programme into another decade.

The A320 is much more than a commercial success because it dared to challenge the established way airliners were built and operated to reshape commercial aviation. It is truly an iconic airliner and the standards it set some 40 years ago raised the bar to establish Airbus as a technological leader.

▲ **LCCs like AirAsia have built their business on the A320.** Airbus

Why the A320?

The first Airbus aircraft, the A300, was launched at the 1969 Paris Air Show and the twin-engined jet operated its maiden flight in October 1972. By May 1974, it had been certified by European and US regulators and in the same month

▲ **Airbus assembles A320 Family jets in four countries.** Airbus

it entered commercial service with Air France.

To meet airline needs the smaller A310 was unveiled in 1978 with launch customers Swissair and Lufthansa starting flights in April 1983. The A310 was produced in a medium range (A310-200) and a longer range (A310-300) with the latter becoming the standard production version. The demise of the series came with the introduction of the A330 in 1994 so even from those early days Airbus was continually looking to the aircraft family model.

Airbus needed an aircraft to break the dominance of Boeing and McDonnell Douglas in the single-aisle market and in June 1977 a new Joint European Transport programme (JET) was established by British Aerospace,

▶ **A320s have been part of easyJet's fleet from the start.** Airbus

▲ **Saudi LCC flynas received the record-breaking A320neo.** flynas

Aerospatiale, Dornier and Fokker. This programme was later transferred to Airbus and settled on three variants, SA1, SA2 and SA3, covering the 125-180 seat markets.

In February 1981, the project was redesignated as the A320 that would carry 150 passengers and was officially launched by Airbus in March 1984. It took three years for the A320 to get into the air with its maiden flight on February 22, 1987 in Toulouse, France, followed by European certification in February 1988.

In March 1988, the first A320 was delivered to launch customer Air France, which had signed a letter of intent for 25 A320s and options for a further 25 at the 1981 Paris Air Show. The first commercial A320 service operated on April 8, 1988, between Paris and Berlin via Dusseldorf.

Other customers to back Airbus and the new aircraft included British Caledonian, Cyprus Airways, Pan Am and Inex Adria Airways. As with the A300, it was a US airline that showed great faith in the A320 when in 1986 Northwest Airlines placed an order for 100 A320s powered by CFM International CFM56-5-A1 engines.

The family expands

To be precise, the name A320 is reserved for the original mid-size aircraft but can often be used to indicate any of the variants, which is why the full array is usually referred to as the A320 Family. With seat capacity ranging from 100 (A318) to 220 (A321), the A320 competed with Boeing's 717, 737 and 757 and all variants have had 180-minute ETOPS granted by EASA in 2004 and the FAA in 2006.

The A320 entered service in 1988 and was followed by the stretched A321 (1994), the shorter A319 (1996) and the shortest A318

▼ **Wizz Air's A321s are a familiar sight in Europe.** Wizz Air

▲ **Air Astana's A321LRs connect Kazakhstan to the world.** Air Astana

▼ **A320neos are ideal for AirAsia's international network.** AirAsia

that entered service with Frontier Airlines in July 2003. Each type is also available as corporate/private jets under the ACJ designation.

All variants have a cruise speed of 448kts and a fuselage height of 13ft 7in, width of 13ft and cabin width of 12ft 2in, with ranges of 3,100nm, 3,750nm, 3,350nm and 3,200nm for the A318/319/320/321, respectively. The A320 is 123ft 3in long while the A318 is 103ft 2in, the A319 is 111ft and the A321 is 146ft long.

Technology breakthroughs

Perhaps the biggest advantage of the A320 was its clean-sheet design, which enabled engineers to develop a totally new way of flying rather than patch new technology on to an existing aircraft. The A320 was officially launched in 1984, whereas the Boeing 737 had its first flight in 1967, so it is unsurprising that the technologies and operating philosophies employed are different.

Despite their differences, both aircraft have been spectacular successes and it is only now, nearly 60 years after its first flight, that the 737 looks constrained by its legacy platform. The comparison is not about which one is better but rather the arrival of the A320 sparked the debate of computerisation and pilot control and forced the industry to reconsider the balance between pilot, aircraft and software.

The most radical element of the A320 was its full fly-by-wire flight control that replaced mechanical and hydraulic linkages with digital systems. It is important to remember the A320 was launched more than 40 years ago and FBW replaced the manual connections that up until then airline pilots had relied on for their entire flying careers.

Airbus countered its critics by pointing out that most accidents stemmed from human error and that intelligent automation, such as the flight envelope protections, would enhance safety rather than shut pilots out of the system. This was

designed to protect the aircraft from excessive bank angles, overspeed, stalls and exceeding structural limits.

The more obvious change was removing the cockpit yoke and replacing it with a side-stick controller, which reinforced the concept of a computer-controlled aircraft with pilots as system managers rather than manual controllers. The side-stick sends electronic signals to the flight control computers and is not mechanically linked between pilots, so it does not move when the other pilot makes an input.

The A320 had a fully integrated glass cockpit with large electronic flight displays replacing analogue gauges and included an Electronic Centralised Aircraft Monitor (ECAM). This system prioritises failures, presents synoptics automatically and displays step-by-step actions for the pilots to follow.

Some of the hydraulically actuated systems on the A320, such as braking, spoilers, flight controls and engine management, were all electronically commanded with thrust settings managed through detents rather than continuous lever positioning.

Retraining and certifying pilots on new types is an expensive and time-consuming process so airlines were attracted to the commonality built into the A320 Family. The four variants share the same flight deck, which means pilots only need one rating to operate all four types – a very significant selling point.

The common rating allowed pilots to transition across the family with minimal training giving fleet planners the flexibility to match aircraft capacity with demand and have pilots on hand to fly any A320 Family aircraft. To quickly change a flight from an A318 to an A321 and use the original scheduled crew gave airlines a major advantage and this commonality soon became a key selling feature for Airbus.

Perfect timing for the neo

Since the introduction of the A320, Airbus has made incremental improvements along the way that kept the airliner current as a modern technological platform. The most

▲ **A321XLRs are reshaping long-haul networks.** Airbus

▼ **Qantas received its first A321XLR in 2025.** Qantas

▲ **The striking livery of an all-black A321neo.** Air New Zealand

significant of those came with the launch of the A320neo in December 2010, although the new derivative did not enter service until January 2016.

The major selling point of the new type was the introduction of new-generation engines from Pratt & Whitney and CFM International that reduced fuel consumption by more than 16%. The neo designation stands for 'new engine option' and it was these powerplants that made the difference, with Airbus adding a few more per cent with new sharklets, weight savings and more efficient aerodynamics.

Airbus was quick to claim the credit for the performance and sustainability gains, but without the promise of the new Pratt & Whitney GTF (geared turbo fan) and the CFM International LEAP-1A engines the A320neo would not have been launched in 2010. Customers usually select the engines and with two options that can sometimes be a lottery as those who chose the Pratt & Whitney GTF would soon discover.

The A320neo had its first flight on September 25, 2014, and powered by Pratt & Whitney PW1100G-JM engines was certified simultaneously by EASA and the FAA on November 24, 2015. Dual certification for the A320 and CFM LEAP-1A combination followed on June 1, 2016, although the A320neo had already entered service with Lufthansa in January 2016.

While both engines faced entry-into-service reliability issues the GTF was the most troubled, with Airbus forced to park airframes in 2017 while Pratt & Whitney sorted out its production issues. More trouble was ahead when some GTF-engined A320s were grounded and delivery schedules were severely affected throughout 2018.

The GTF and LEAP engines have been extremely successful in reducing fuel burn and harmful emissions, but paradoxically both have failed when it comes to their durability with maintenance events occurring much earlier than planned. As the calendar flipped over to 2026 Pratt & Whitney was still dealing with GTF repair and maintenance issues although new-build engines are now performing perfectly.

Neo variants

The success of the broader A320 Family meant that new engine options would also be available for the variants, with the A321neo and the A319neo forecast to arrive within a year of the A320's entry into service. The A321neo entered production and had its first flight in 2016 and by March 2017 it was certified by EASA/FAA for both engine types.

Virgin America received the first A321neo and it entered service, fitted with CFM LEAP engines, on May 31, 2017. In addition to the new engines it differed from the A321ceo ('current engine option') with structural strengthening, redesigned wing and undercarriage and extended fuselage fuel tanks that increased the range to 4,000nm and MTOW to 97 tonnes.

In 2015, Airbus launched the A321LR that increased the range to 4,500nm with an initial two-class layout of 206 seats. The LR added three Additional Centre Tanks (ACTs) to give it more range than Boeing's 757, opening up transatlantic markets. It was certified in 2018 with approval for ETOPS up to 180 minutes and entered service with Israeli airline Arkia on routes to London, Paris, Barcelona, Zanzibar and the Seychelles.

A321LRs have proven very popular on medium-haul routes across the Atlantic, such as JetBlue's services from Boston and New York to London which began in August 2021. Australian LCC Jetstar had used the LR to run domestic services during the day and then head to international destinations, such as Bali and Fiji overnight before resuming domestic flights the next morning.

At the 2019 Paris Air Show, Airbus set the exhibition alight with its official launch of the A321XLR, securing orders and commitments for more than 200 of the new type that would not enter service until late 2024. The first flight of the XLR was in June 2022 with certification using the CFM LEAP 1A in July 2024 and with the Pratt & Whitney GTF in February 2025.

With a flight time of up to

▼ **Airbus A320s operate in all parts of the world.** Royal Jordanian Airlines

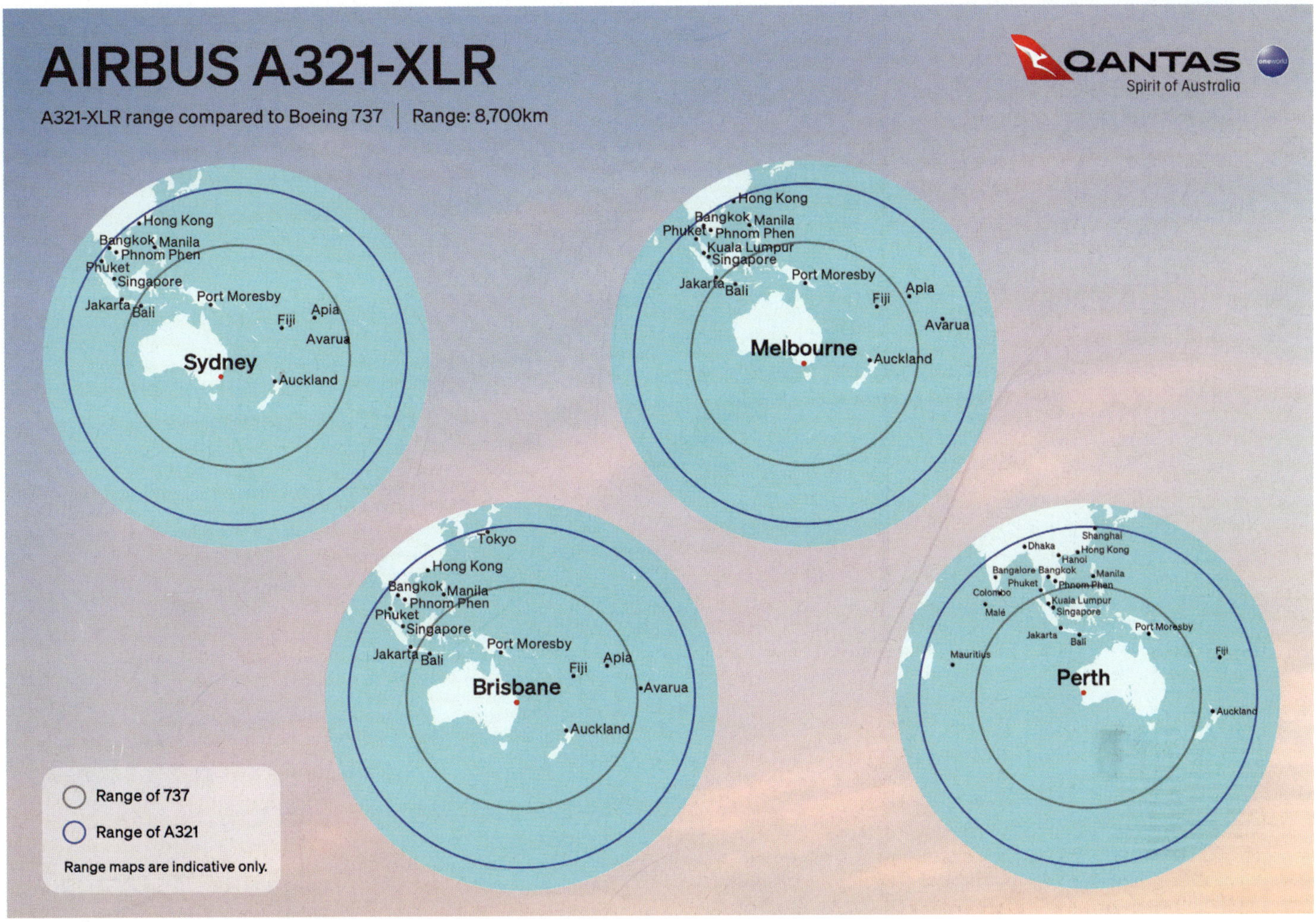

▲ **A321XLRs have a clear range advantage over 737s.** Qantas

11 hours and range up to 4,700nm, the XLR can connect destinations as diverse as New York to Rome, London to Vancouver, Delhi to London and Sydney to Kuala Lumpur. These routes are usually the preserve of widebody aircraft but airlines can now operate them with the lower operational costs of a narrowbody.

The first was delivered to Spanish carrier Iberia configured with 182 seats in a two-class layout with lie-flat business class seats and direct aisle access. With in-seat connectivity for all passengers and 60% larger overhead luggage capacity, the XLR has the feel and comfort of a widebody making it an ideal aircraft for launching thinner medium-haul routes.

The extended range is possible by installing a permanent Rear Centre Tank (RCT) that holds 3,400 US gallons of fuel and supersedes the modular ACTs employed on the A321LR. The RCT is integrated into the aircraft's airframe and Airbus also reinforced the main landing gear to structurally handle the higher MTOW of 101 tonnes.

A family affair

With minimum disruption and much lower development cost, Airbus has reinvented the A320 and launched a suite of new and improved offerings for airlines of all shapes and sizes. Both full service and low-cost carriers can build their fleets with A320neos to open new longer and thinner routes or to add frequency on existing services where demand fluctuates.

The A321XLR can work alone or in tandem with a widebody on routes with the narrowbody flying on lower demand days and the larger aircraft stepping in to take the load for high demand slots.

Today's A320 is not the same aircraft it was in 1988 and thanks to the foresight of the fledging Airbus team it has stayed relevant for nearly 50 years since it was conceived in 1977. The Airbus A320 has changed the way airlines operate and the A321XLR is opening up opportunities for airlines to fly longer distances with less cost, which surely puts it in the 'iconic' class.

▼ **Air Transit's A321LR departing Toulouse.** Airbus

TRIPLE SEVEN Keeps Going

The first 777 was delivered more than 30 years ago, but with its 777X successor floundering it will be flying for a while longer.

▲ **Air New Zealand flies 777-300ERs on long-haul services.** Air New Zealand

▼ **Emirates has around 140 Triple Sevens.** Emirates

The Boeing 777 is iconic for many reasons and with 1,774 delivered between 1995 and 2024 it is highly likely that most readers will have travelled on the widebody. It was introduced to service in June 1995 with United Airlines and the final 777 was rescued from storage and entered service with Ethiopian Airlines, via lessor Altavair in December 2024.

Why the 777?

When Boeing launched the 777 in 1990, the industry was in another period of change with twin-engined jets like the A310 and 767 or trijets such as the L-1011 and DC-10 filling the gap in the market between narrowbody aircraft and the 747 jumbo. Changes to ETOPS rules were making long-haul twin-engined aircraft a viable option on transoceanic routes that were once the preserve of three or four-engined airliners.

In response to customer feedback, Boeing saw an opportunity for an aircraft that combined long-range capability with lower operating costs and greater flexibility and that could carry 300-400 passengers in comfort. In some respects the Airbus A330 was moving into that space, but Boeing aimed to overtake it with a larger, more capable aircraft that could be a better choice than the 747 or DC-10 on lower density long-haul routes.

As it had done with the 747, Boeing consulted deeply with a panel of airline customers who would find the concept attractive – including United Airlines, Cathay Pacific, British Airways, Japan Airlines, American

Airlines, Qantas, Delta Air Lines and All Nippon Airways – that became known as the Working Together Group.

When the group first met in January 1990 they were presented with a detailed questionnaire asking what they wanted the aircraft to include and by March the group had decided on a baseline configuration. The airlines wanted a cabin cross-section close to the 747s, capacity up to 325 passengers, flexible interiors, fly-by-wire controls, glass cockpit and 10% better seat-mile costs than the A330 or MD-11.

In October 1990, launch customer United Airlines placed an order for 34 aircraft with options for a further 34 to replace its ageing DC-10 fleet. United specified the 777 had to be capable of flying Chicago to Hawaii, Chicago to Europe and nonstop from Denver's hot and high airport to Hawaii.

▲ **Cathay Pacific operates a three-class layout of 368 passengers.** Boeing

Road to production

While the Triple Seven (777) was a game changer for airlines, it also revolutionised how commercial aircraft were designed, with the new widebody the first to be developed using an entirely computer-aided design (CAD) process. This digitisation eliminated the need for full-scale physical mock-ups, although, just to be sure, Boeing did build a mock-up of the nose section to verify the CAD results, which proved the system.

The system allowed engineers to virtually build the aircraft and ensure that the more than three million parts all met design parameters. The high performance Integrated Visualisation Tool (IVT) was used to support large-scale collaborative design reviews, production illustrations and other uses of the CAD data outside of engineering.

The 777 was assembled in the Everett plant with work starting on January 4, 1993 and on April 4, 1994 the first Triple Seven was rolled out for its initial flight on June 12, 1994. Nine 777s were used in an 11-month flight test programme that included environments as diverse as the Californian desert to the freezing depths of Alaska.

In the same vein as many other aviation gains, it was the engines that drove the technical advancement of the Triple Seven with three OEMs offering powerful turbofans that delivered between 71,000lbf and 115,300lbf. Initial variants gave operators engine choices from Pratt & Whitney, General Electric and Rolls-Royce, but the longer-range 200LR and 300ER offered only the GE90 powerplant.

To meet ETOPS regulations, eight 180-minute single-engine test flights were operated and on April 19, 1995 the 777 was certified by the FAA and Europe's Joint Aviation

▼ **Singapore Airlines has a long history with 777s.** Boeing

▲ **Boeing has worked through various 777 production issues.** Boeing

Authorities, the forerunner of the European Union Aviation Safety Authority (EASA).

The FAA issued 180-minute ETOPS clearance for the Pratt & Whitney PW4084-engined 777, marking it as the first aircraft to have ETOPS-180 at entry into service. That was extended for the longer-range 777-200LR and 777-300ER types and in 2011 regulators granted ETOPS-330 for these variants.

▶ **Garuda Indonesia built a network around its 777s.** Boeing

Family growth

The 777 had immediate success with more than 200 orders and commitments secured before production began in 1993 and by June 1997 that had grown to 323 orders from 25 airlines. The final 777 was delivered in 2024 and by then Boeing had delivered 1,774 aircraft – 571 of the 777-200 family, 893 of the 777-300 family and 310 of the 777F freighter aircraft.

The 777-200 first flew on June 12, 1994 and was delivered to United Airlines in May 1995, entering service on June 7, 1995 with a flight from London Heathrow to Dulles International Airport near Washington DC. This aircraft was powered by Pratt & Whitney engines with the GE90 debuting on a British Airways aircraft in 1995 and the Trent 800 on a Thai Airways International 777 in 1996.

The 777-200, -200ER and -300 are considered the Classics or first generation of the widebody while the 777-200LR, -300ER and 300F are the second generation. Apart from the engines the major difference between the generations is that the second generation 777s have a wingspan of 212ft 7in compared with 199ft 11in on the first and significantly higher MTOWs.

The -200 series was produced with three variants, the -200 aimed at domestic markets, the -200ER for transoceanic routes and the -200LR as its longest range version.

When powered by GE engines in a typical three-class layout, the -200 has an MTOW of 545,000lb and can carry 305 passengers with a range of 5,175nm.

The Boeing specifications show that the -200ER has an MTOW of 656,000lb and can carry 301 passengers with a range of 7,225nm while the -200LR has an MTOW of 766,000lb flying the same number of passengers up to 9,395nm. Deliveries by type were 88 (-200), 422 (-200ER) and 61 (-200LR) and all 777 variants are operated with a flight crew of two pilots.

Seeing more opportunities in the widebody market, Boeing decided to stretch the -200 and launch it as the 777-300 with capacity for 368 passengers in a typical three-class configuration. With a length of 242ft 4in the 777-300 was 33ft 3in longer than the baseline -200 series and with a design range of 6,030nm the larger aircraft had an 855nm range advantage.

The -300 was launched at the Paris Air Show on June 26, 1995 with the first off the production line in September and operating its first flight on October 16, 1995. Eight customers ordered 60 of the 777-300s with 42 opting for the Rolls-Royce Trent 800 and 18 for the Pratt & Whitney PW4000 producing around 98,000lbf.

Gen two arrives

From the start of the 777 programme, Boeing had planned for an ultra-long-range variant and by 1995 its directors had approved specifications for the proposed 777-200X and -300X aircraft. It rejected a proposal to reduce capacity to increase range and called on its three engine suppliers to submit proposals for engines in the 100,000lbf range.

◀ **Emirates is refurbishing 777s while waiting for its 777Xs.** Emirates

▼ **Refurbishing Emirates 777 cabins has been a costly project.** Emirates

▲ **Korean Air operates the 777-300ER.**
Korean Air

General Electric proposed to develop the GE90-115BL engine delivering 115,000lbf and in 1999 Boeing selected this engine as the sole powerplant for the second generation –300ER and -200LR aircraft.

Boeing's 777-300ER was the first second-generation aircraft built and it made its initial flight with the FAA and EASA granting certification on March 16, 2004. As the launch customer, Air France received the first 777-300ER on April 29, 2004 and it became the most popular 777 type with 838 delivered to customers globally.

The arrival of the-300ER gave airlines a high-capacity, long-range, twin-engined airliner that could replace four-engined aircraft and deliver significantly lower operating costs. Its powerful GE-90-115B engines and extended range of up to 7,930nm combined with an exit limit of 550 passengers gave airlines great scope and flexibility to match seats to demand on specific routes.

Middle Eastern full service carrier Emirates operates the largest fleet of 777s with a mix of around 140 aircraft, including 777-300ER and 777-200LR variants. The -200LRs have a two-class layout of 302 seats configured for 38 business and 264 economy passengers while the 777-300ERs range in seats from 320 in a four-class layout to 421 in a two-class configuration.

The most common iteration on the Emirates -300ER is a three-class layout of 354 passengers with eight in first, 42 in business and 304 in the main cabin. The airline is progressively introducing a premium economy section, which is first being fitted on aircraft dedicated to ultra-long-haul routes (ULR), such as to Australia and the Americas.

In this ULR four-class layout, the -300ER typically carries up to 332 passengers – eight in first, 40 in business, 24 in premium and 260 economy class passengers. Emirates has at least five -300ER cabin configurations, which gives it the flexibility to switch aircraft around when seasonal or peak demand fluctuates without impacting crewing schedules.

In Asia, Singapore Airlines has a long history with the 777 and in early 2026 has approximately 22 -300ERs in service, which it operates alongside Airbus A380 and A350 and Boeing 787 and 737-8 MAX passenger aircraft. Its 777-300ERs are configured in a four-class cabin of 264 seats – four in first, 48 in business, 28 in premium economy and 184 in economy.

The -300ER also paved the way for more passenger comfort, with its wider cabin and aisle widths allowing for the introduction of lie-flat business class seats, premium economy

▼ **Emirates uses 777s to open new services globally.**
Emirates

▲ **Boeing is gearing up to produce the 777X.** Boeing

cabins and high density but comfortable economy cabin layouts. The extra space facilitated larger overhead bins and the quieter engines and improved cabin architecture provided passengers with a smoother and quieter ride.

The smaller 777-200 series was not left behind with Boeing developing the -200LR Worldliner, so named because of its ability to connect almost any two airports in the world. In a typical two-class layout it was configured with 317 seats and had an exit limit of 440 passengers.

The first -200LR was delivered to Pakistan International Airlines on February 26, 2006 and a total of 61 were delivered to 12 operators as a direct competitor to the four-engined Airbus A340 quadjet and the A350ULR used on routes such as Singapore-New York.

▶ **Airlines are waiting impatiently for the new 777X.**
Boeing

It was by far the longest range 777 with the Boeing specifications listing a design range of up to 9,395nm when fitted with auxiliary fuel tanks, compared with the -200ER's range of 7,725nm and the -300ER's of 7,939nm. Each auxiliary tank holds 1,875 US gallons and fitting the aircraft with three brings the total fuel capacity to 53,515 US gal compared with 31,000 US gal on the 777-200 and 45,220 on the 777-200ER.

On November 10, 2005 the first -200LR backed up Boeing's Worldliner title when it set a (then) record for the longest nonstop flight of a passenger airliner. The 777-200LR flew eastward from Hong Kong to London covering 11,664nm, well beyond its design range, on a flight lasting 22hrs 41mins, which was logged in the *Guinness World Records*.

777F completes the family

The -200LR operated with the same GE90-115BL engines that powered the -300ER and had a MTOW of 766,000lb. The -200LR provided the platform for the 777 Freighter (777F) that had its first flight on July 14, 2008 with the two sharing the same structural design, airframe, engines and fuel capacity.

The 777F was jointly certified by the FAA and EASA on February 6, 2009 and launch customer Air France received its first 777 freighter on February 19, 2009. It quickly became a favourite of air cargo operators as its maximum payload of 224,900lb was close to the 747-200F maximum of 244,900lb but with reduced operating costs.

Boeing saw 777F as an ideal replacement for older and less fuel-efficient freighters such as the 747F, MD-11F and Airbus A300/A310 aircraft, particularly on intercontinental routes. The 777F benefitted from the GE-90 engines and additional fuel tanks to have a range of up to 9,750nm when lightly loaded or 4,970nm at its maximum structural payload.

The main deck can hold 27 standard pallets with additional containers in the lower holds, and the large cargo side door and reinforced floor are optimised for high density freight loads. The long range of the 777F enabled nonstop freight services at maximum revenue payload from Hong Kong to Sydney, Istanbul and Anchorage or from Dubai to Perth, Tokyo, Dakar and Hong Kong.

Where is the 777X?

The long-awaited and much hyped 777X is somewhere over the horizon but can't be included in a catalogue of iconic airliners, simply because at the time of writing in early 2026 it had not even made it to certification.

The new type has been successful in gaining orders and commitments for more than 600 aircraft while already being five years late and with no firm date set by Boeing for its entry into service. The 777X programme was launched in November 2013 based on orders from Emirates, Etihad Airways, Lufthansa and Qatar Airways, with the aircraft projected to enter service in 2020.

The first prototype, a 777-9, emerged from the factory in March 2019 and operated its maiden flight on January 25, 2020. By then Boeing had already pushed first deliveries back to 2022/2023 and then to 2025/2026 but at the end of October 2025 it was delayed further to 2027, seven years later than originally announced.

According to Boeing's November 2025 Orders and Deliveries report, the 777X had gathered orders from 15 customers for 619 aircraft, with two aircraft ordered by an unidentified customer or customers. The customer list is an honour board of some of the world's most prestigious airlines, all of whom are waiting for aircraft to replace ageing and less-efficient aircraft in their fleets.

▼ **Air China has operated 777s on a variety of routes.**
Boeing

A staggering 44% of those 777Xs have been ordered by Emirates, which is now spending billions of dollars refurbishing 777s that it had expected to have retired and taken out of service years ago. Emirates is not alone in doing this, but with 270 on order its leadership is the most vocal about the many delays and failure in the project that it is now paying for.

Apart from Emirates some of the other airlines and their orders include Qatar Airways (124), Cathay Pacific (35), Singapore Airlines (31), Lufthansa (27), British Airways (24), China Airlines (23), Korean Air (20), All Nippon Airways (17), Etihad Airways (15) and Air India (10), with Cargolux Airlines and Silk Way West Airlines waiting for the new 777F freighter.

777's legacy

The first half of the 2020s has been a disaster for Boeing, but for more than 30 years the 777 family of widebodies has delivered exceptional service to billions of passengers around the world. The aircraft revolutionised long-haul travel with its spacious widebody cabin, reliability, range, efficiency and versatility as either a passenger or cargo aircraft.

It proved an ideal replacement for the previous-generation 747 and A340 fleets by delivering similar capacity but with just two significantly more efficient engines its operating economics lined the pockets of airlines for decades. Boeing challenged engine OEMs to develop more powerful and cleaner engines and even after all these years the iconic 777 is still the benchmark new entrants are measured against.

▼ **Boeing has more than 600 orders for the 777X.** Boeing

When Size Matters

When Airbus needed an aircraft to challenge the 747, it decided to go all-in with a double-decker that could carry 853 passengers.

In today's usage, the word 'iconic' can have multiple meanings, but a scan of dictionaries settles on someone or something that is "widely known and acknowledged especially for distinctive excellence".

For this publication, the benchmark for iconic status is aircraft that have significantly changed the course of commercial aviation rather than simply doing what already exists a little bit better. The Airbus A380 ticks the box and despite all the criticisms surrounding the programme it lives on as the preferred choice for millions of long-haul travellers.

After the success of the 747 with its upper deck it is no great surprise that the big aircraft OEMs and airlines saw potential in a full-length double-decker airliner, but it was Airbus that had the courage to build it and the confidence of Emirates to heavily support it that brought the super jumbo to life.

While the 747 is widely venerated as an outstanding success, its closest cousin, the Airbus A380, divides opinion about how successful the aircraft was and the legacy it left behind. However, critics and fans alike would agree the A380 brought its own special magic to aviation with the idea that an aircraft could move more than 800 people across continents in style and comfort.

Some observers point to the huge financial investment of around £20-£25bn for the sale of just 251 aircraft and short production span as proof of the A380's failure, while others highlight its success in developing long-haul markets, especially through the Middle East, and the programme's innovative technology that is now seen on modern airliners such as the A350 and A321XLR as the legacy.

Success and failure can be measured in many different ways, but what is not in dispute is the technological achievement of

▼ **Qantas returned its A380s to service after the COVID-19 pandemic.** Qantas

Airbus to design, manufacture and introduce the A380 that started more than 30 years ago. Even if only for that reason alone, the double-decker has earned its place in this catalogue of airliner icons.

Combatting the 747

By the late 1980s, Airbus was gaining in experience and confidence following the successful introduction of its initial offering, the A300 series and development of the four-engined A340 and twinjet A330. The A340-300 entered service in March 1993 and could carry 250-290 passengers with a range of 6,700nm, while the A340-500 could fly 310 passengers up to 9,000nm, making it the longest-range aircraft of its time.

With the A340, Airbus was a serious contender in the long-haul, high-capacity market, but its ambitions were set a lot higher than that. In 1988, it began studies on the development of an ultra-high-capacity airliner to boost its portfolio

▲ **The A380 is huge from any angle.** Airbus

▼ **Emirates operated nearly half of all A380s.** Emirates

▲ **Singapore Airlines launched the A380 in 2007.** Airbus

▼ **Airbus manufactured the A380 in France.** Airbus

and break the dominance of Boeing's 747 in this market segment.

Those studies progressed and, after flirting with joint ventures, Airbus announced in June 1994 that it would go it alone and develop its own very large airliner, the double-decker A3XX. Airbus believed the A380 was ideal for the prevailing hub-and-spoke network model, while Boeing saw the market shifting to more point-to-point routes that would be better served by its 777 widebody.

After navigating the perils of the 1997 Asian financial crisis, Airbus launched the now-designated A380 project on December 19, 2000, backed up by orders for 50 aircraft from six launch customers. The launch customers for the A380 were Singapore Airlines, Emirates, Qantas, Air France, Lufthansa and Korean Air with Singapore Airlines receiving the first aircraft on October 15, 2007.

After the launch, progress moved quickly with the first A380 unveiled in Toulouse on January 18, 2005 and operating its maiden flight on April 27. Airbus produced five aircraft for testing and demonstration, which included high-altitude flights in Addis Abada, transatlantic testing between Europe and Colombia and cold weather testing in Nunavut in Canada.

In March 2006, aircraft evacuation was tested and with eight of 16 exits randomly blocked, 853 passengers and 20 crew exited the darkened aircraft in 78 seconds, well inside the 90 seconds required for certification. Following that, the A380 was approved to carry up to 853 passengers and on December 12, 2006 it was granted EASA and FAA joint certification.

Superjumbo takes off

Nicknamed 'Superjumbo', the A380 entered commercial service on October 25, 2007 operating Singapore Airlines flight SQ380 from Singapore to Sydney, Australia. The second A380 was delivered to Emirates and launched on the Dubai-New York route in August 2008, followed by a delivery to Qantas that entered service between Melbourne and Los Angeles in October of the same year.

Air France received its first A380 in October 2009 and Lufthansa in May 2010 and by July 2010 the 31 A380s then in service had carried six million passengers on 17,000 flights between 20 international destinations. Those numbers grew exponentially with 100 million passengers carried by September 2015 and by December 2021 the A380 fleet had carried more than 300 million people on more than 800,000 flights.

Along the way, Airbus encountered significant production delays, especially given the 330 miles of wiring in each aircraft and the availability of Rolls-Royce engines. These delays led to Airbus rescheduling delivery schedules and compensating 13 customers, including Emirates, Singapore Airlines, Qantas, Air France, Qatar Airways and Korean Air.

Passenger experience

Initially, Airbus had envisioned the A380 would carry 555 passengers in a three-class layout or 853 passengers in a single-class economy configuration, with 538 on the main deck and 315 upstairs. In 2007, it offered a configuration carrying 525 passengers in a three-class layout that Airbus said would add an additional 200nm to the range.

Finding an aircraft's range is a bit like asking for a piece of string in that range depends on variables, such as the cabin configuration, MTOW, fuel capacity and other factors that can differ from aircraft to aircraft. In its January 2022 Facts & Figures, Airbus lists the range of the A380-800 in a standard four-class, 545-seat configuration as 8,000nm with an MTOW of 560 tonnes and fuel capacity of 320,000lit (84,535 US gal).

Early operators typically configured their aircraft with around 490 seats in a three-class layout, although Korean Air opted for a roomier 407 passengers compared with Emirates' layout of 615 seats in a two-class layout. Within their fleets, airlines often had a variety of cabin configurations to cater for the routes A380s operated, such as ultra-long-haul, medium-haul or high-density domestic routes.

Bearing in mind that Airbus positioned the A380 as direct competition to the 747, the European OEM highlighted these differences, such as the A380 having 5,920sq ft of usable floor space or 40% more than the next largest airliner, the 747-8. Airbus also said that the A380 had 50% less cabin noise, 50% more cabin area and volume, larger windows, bigger overhead bins and 2ft more headroom than the 747-8.

Airlines have typically spent the most on designing lavish premium cabins and as far back as 2007, launch customer Singapore Airlines had introduced partly enclosed first class suites. The bar lounge area featured on Emirates, Etihad and Qatar Airways aircraft is perhaps the most photographed part of the A380, although the shower and spa available to Emirates' first-class passengers is the epitome of luxury.

The COVID-19 pandemic accelerated the retirement of many A380s, but when aviation reopened the rush for more capacity pushed many airlines to get their A380s out of desert storage and back into operation. Carriers such as Qantas, Emirates and Singapore Airlines are making multi-million dollar investments on refurbishing cabins as they position their A380s for another decade of service.

QF32 setback

The pace of the rollout and deliveries received a setback on November 4, 2010 when a Qantas A380 taking off from Singapore Changi Airport suffered an uncontained failure of the number two engine. As the aircraft, registered VH-OQA, climbed through 7,000ft the flight crew heard two bangs, which the accident report said were the result of an uncontained failure of the No 2 Rolls-Royce Trent 900 series engine.

Debris from the engine impacted the aircraft causing significant damage to the structure and aircraft systems and resulted in fuel leakage from the left wing fuel tank. After circling for 50 minutes to assess the health of the aircraft's systems and performing progressive aircraft controllability checks, the flight crew landed the aircraft safely at Changi Airport.

After landing the No 1 engine could not be shut down and with hot brakes and fuel leaking nearby the crew delayed offloading passengers for approximately one hour. With emergency services on hand, the 420 passengers and 29 crew later left the aircraft with no injuries reported, although it took authorities three hours to shut down the engine by spraying firefighting agent into the engine inlet.

The Australian Transport Safety Bureau investigation determined that the probable cause of the engine failure was a manufacturing error involving an internal oil feed stub pipe. The investigation report said the oil feed pipe was manufactured with reduced wall thicknesses that eventually cracked and caused oil spray leakage and fire in the High Pressure/Intermediate Pressure (HP/IP) bearing support hub assembly.

The ensuing fire weakened the intermediate turbine drive arm, which subsequently fractured, allowing the IP turbine disk to overspeed and break apart and high speed debris to penetrate many areas of the A380.

In June 2011, Qantas announced that it had agreed to compensation of £62m from Rolls-Royce and the aircraft was repaired at an estimated cost of £70m before returning to service in April 2012.

During the repairs, cracks were discovered in wing fittings and as a result in January 2012 EASA issued an airworthiness directive that all A380s with under 1,800 flight hours were to be inspected within six weeks or 84 flights, while those with more than 1,800 flight hours were to be examined within four days or 14 flights. On February 8, 2012 the checks were extended to cover all 68 A380s then in service, with Airbus paying estimated costs of €200m (£174m).

◀ **Emirates is operating its A380s into the 2030s.** Emirates

A380 airlines

The first A380 was delivered to Singapore Airlines in 2007 and the 251st and last was handed over to Emirates in 2021. Over those 14 years the A380 was operated by 15 airlines and by the start of 2026 ten carriers were still operating their Superjumbos.

The airlines operating A380s in January 2026 were Emirates, Singapore Airlines, Qantas, British Airways, Lufthansa, Qatar Airways, Korean Air, Asiana Airlines, Etihad Airways and All Nippon Airways. Former operators included Air France, Thai Airways, Malaysia Airlines, China Southern Airlines and wet-lease specialist Hi Fly.

Emirates is by far the largest A380 operator taking 123 aircraft out of the 251 delivered and without the airline's unwavering support it is hard to see how the programme would ever have got

▼ **British Airways opted for Rolls-Royce Trent engines.** British Airways

▲ **Lufthansa's A380s remain in service in 2026.** Airbus

off the ground. The second largest user was Singapore Airlines (24), followed by Lufthansa (14), British Airways (12), Qantas (12), Etihad Airways (10), Korean Air (10), Air France (10) and Qatar Airways (10), with all others in single digits.

Between them, the large Middle Eastern carriers – Emirates, Etihad Airways and Qatar Airways – took delivery of 57% of all A380s with the aircraft an ideal choice for the hub-and-spoke models they operated. This was exactly what Airbus had envisioned for the A380 with long-haul passengers flown to hubs in Dubai, Doha and Abu Dhabi and then onwards to their final destination.

With relatively small local populations, this was the only way for these national airlines to grow market share and the size and economics of the A380 perfectly suited their operational and aspirational objectives.

After forging a partnership with Emirates, Qantas shifted its transfer hubs from Singapore and Hong Kong to Dubai using A380s flown from Australia. That changed when Qantas started operating nonstop 787 flights from Perth to London in 2018 and returned to transiting the A380s through its former Asian hubs.

▼ **In 2021, Emirates welcomed its 123rd and final A380.** Emirates

Lufthansa, Korean Air, Singapore Airlines and British Airways favoured the Superjumbo on transcontinental long-haul routes, while China Southern Airlines used the A380 to great effect on its busiest high-capacity domestic routes, such as Shanghai-Beijing, as well as for flights to Australia, Europe and North America.

The A380 proved itself with airlines operating both hub-and-spoke or point-to-point routes where high capacity is needed rather than more frequency with smaller aircraft. In 2019, All Nippon Airways began nonstop flights with its three A380s from Tokyo to Honolulu where the seating capacity suited the high volumes of tourists who flock between the two destinations.

An inglorious end

At the 2016 Farnborough Airshow, Airbus said that starting in 2018

▲ **Charter operator Hi Fly operated one A380 in 2018-2020.** Hi Fly

it expected to deliver 12 A380s per year, down from the 27 it had delivered in 2015. With sales and profitability declining, a possible order for 36 A380s from Emirates would have ensured production beyond 2020, but Emirates wanted assurances that A380 production would be extended until 2028.

The pair didn't find a way through that and on February 14, 2019, Emirates cancelled its order for 39 Superjumbos and replaced it with orders for 40 A330-900s and 30 A350-900s. In return, Airbus said that this cancellation would bring A380 production to an end when the last unfilled orders were delivered in 2021.

At the time, Airbus still had 17 more A380s to complete – 14 for Emirates, three for All Nippon Airways, with the first of the three delivered on March 21, 2019. The last A380 was delivered to Emirates in December 2021 and the curtain fell on A380 production, although the aircraft will be flying well into the 2030s.

Success or failure?

Many of the features seen on modern aircraft like the A330neo and A350 first saw service on the A380 and perhaps that needs to be remembered when pundits criticise the development cost overruns attributed solely to the A380 programme.

◀ **Qantas has refurbished its A380 cabins.** Qantas

▶ **British Airways operated 12 A380s on long-haul routes.** British Airways

▼ **A row of A380s at Dubai Airport had been a familiar sight since 2008.** Emirates

The A380 was made for nonstop long-haul flying by a flight crew of two, so Airbus eased pilot workloads with new cockpit technology that enhanced its fly-by-wire systems. On the A380, Airbus introduced modular avionics, head-up displays, integrated the traffic avoidance and collision avoidance system into the autopilot and developed an automated brake and vacate feature for landing.

The Superjumbo looked after passengers by introducing the widest economy seat, quietest cabin, freshest air, variable mood lighting and groundbreaking inflight entertainment systems that have made flying today so much more comfortable.

◀ **Emirates has invested heavily to keep its A380s current.** Emirates

Perhaps the question was answered by some of the world's leading airlines that quickly reactivated parked A380s when demand unexpectedly surged back after the pandemic and capacity was in short supply. Many of those airlines had virtually retired their A380s, but when the chips were down the double-decker shone through.

When viewed in isolation the A380 may not have lived up to its financial objectives, but there is no doubt it has changed the face of aviation and its DNA lives on in those aircraft that have followed. How may aircraft have left such an enduring legacy?

Tomorrow's Icons

Commercial aviation never stands still and, as our airliner icons age, which aircraft are going to fill their shoes?

It takes decades to launch, design, test and bring into service a new clean-sheet design aircraft and as 2026 opens both Airbus and Boeing have their attention on a new mid-range aircraft to take over from the A320 and 737 programmes sometime in the 2030s.

Those two aircraft are the workhorses of commercial aviation and with sustainability targets and the commitment to net zero emissions by 2050 looming, the OEMs have to innovate with new airframes and clean propulsion technologies to give airlines the tools they need to meet those goals.

The 15 aircraft we have reviewed are all legitimate icons because each changed the course of aviation and left their own stamp on air travel. Commercial aviation is a thriving ecosystem where lessons are learned that inform the future and today's innovations are the platform for tomorrow's breakthroughs.

To bookend this icons collection it is worthwhile having a look at which of the latest generation aircraft not already covered will be the icons of tomorrow and what they are doing now to earn that status. Three showing that promise are the Airbus A350, Boeing 787 and Embraer's E2 family of jets.

Ultra-long-ranger

In early 2004, Boeing launched the 787 Dreamliner programme that Airbus initially believed would be a competitor to the mid-range A330, but when orders started flowing for the new aircraft Airbus realised it was losing ground in the two-engine widebody market and needed to react.

Airbus first proposed a derivative of the A330 featuring improved aerodynamics with composite wings and advanced winglets and new-generation engines similar to those offered on the 787. Airlines greeted that with little enthusiasm and after a stuttering start Airbus announced the new, clean-sheet A350 XWB (xtra wide body) at the 2006 Farnborough International Airshow.

The XWB designator was used because the A350 has a 5in wider fuselage than the 787 that allowed it to offer up to ten abreast seating, compared with the 787's eight, with a maximum seating capacity of 440-475 depending on the variant. Initially, Airbus proposed three variants, A350-800, -900 and -1000, although the shorter -800 was later cancelled with customers opting for the larger A350-900.

The first A350-900 entered service with launch customer Qatar Airways on January 15, 2015 and the A350-1000 was first operated by Qatar Airways from Doha to London on February 24, 2018. The success of the A350 marked a pivot for Airbus as it shifted focus from the A380 to the next generation of twin-engined widebodies more suited to the resurgence of point-to-point networks.

With a length of 242ft 1in, the A350-1000 is 23ft longer than the -900 and with a considerably higher MTOW (710,000lb) and more powerful engines it attracted the attention

▲ **Project Sunrise connects Australia to London and New York nonstop.** Qantas

▲ **First Qantas A350-1000ULR rolls out for testing in 2026.** Qantas

of transpacific airlines such as Cathay Pacific, Singapore Airlines and United Airlines. Its capacity and range of 9,000nm made the -1000 a potent competitor to the 777-300ER while the -900 was building its presence in markets captured by the 787.

Singapore Airlines operated a specially configured A350-900ULR on the 8,300nm Singapore-Newark (US) route and A350s were popping up on other ultra-long-range services such as Manila-New York, Singapore-New York and Singapore-San Francisco.

▶ **Qantas A350-1000ULR first class suite.** Qantas

▼ **Qatar Airways launched both the A350-1000 and A350-900.** Airbus

For more than a decade, Australia's Qantas has talked of connecting East Coast Australia to London or New York with nonstop flights and in May 2022 the airline ordered 12 A350-1000ULRs specially modified for its Project Sunrise. The modifications included an additional rear centre fuel tank and associated systems to enable flights up to 22 hours, bringing any global destination within reach on a nonstop flight.

In November 2025, the first aircraft rolled off the fuselage assembly line and into the next manufacturing stage before undergoing an extensive testing programme in 2026. This is the first of 12 and is expected to be handed over to Qantas in late 2026, with the first commercial Project Sunrise services starting in the first half of 2027.

Less than a decade after the first A350-1000 entered service, the aircraft will soon set new records for ultra-long-haul commercial services and bring Australia within easy reach of any city in the world, as distinct from lengthy stopovers in

▲ **Qantas has ordered 12 bespoke A350-1000ULRs.** Qantas

◀ **Composite materials reduced weight on 787s.** Boeing

Asia or North America. That should be enough for the A350 to claim a spot the next time a collection of airliner icons is compiled.

Dreamliner's time

Boeing's 787 Dreamliner programme was launched in 2004 and first flew in December 2009 before entering service with All Nippon Airways (ANA) on October 26, 2011. The twin-engined widebody was designed to carry 200-300 passengers up to 8,500nm on point-to-point routes to meet the changing network demand away from hub-and-spoke models.

Its arrival was not without growing pains and in response to on-board fires caused by lithium-ion batteries, the FAA grounded all 787s in January 2013 until a revised battery design was approved in April 2013. Deliveries were later interrupted in 2019 due to quality control lapses and from January 2021 until August 2022 almost no Dreamliners reached customers.

The 787 hastened the shift from the hub-and-spoke models that were the lifeblood of four-engined widebodies like the A380 and 747 by reshaping airline economics, passenger expectations and how aircraft were designed and manufactured. Boeing outsourced much of the design work to the manufacturers who supplied the components and, despite some hiccups and cost overruns, built an effective global supply chain.

The step change came with the widespread adoption of composite materials to reduce weight, with around 50% (by weight) of the 787's primary structure made from carbon fibre reinforced plastic. This produced a lighter, stronger airframe that is more resistant to fatigue and corrosion compared with fuselages typically made with aluminium sheeting held together by more than 50,000 fasteners.

At 186ft long, the 787-8 was the first variant to fly in December 2009 followed by the longer 787-9 (206ft) in September 2013 and the longest 787-10 (224ft) in March 2017. Boeing lists the -8 carrying 248 passengers up to 7,307nm, the -9 carrying 296 passengers up to 7,565nm and the -10 seating 336 passengers up to 6,330nm.

The timing of the 787's development allowed it to offer

▼ **Singapore Airlines was an early 787 operator.** Singapore Airlines

▲ **Qantas nonstop Sydney-London starts in 2027.** Qantas

new-generation and highly efficient engines with the choice of either the General Electric GEnx-1B or the Rolls-Royce Trent 1000. Both engines delivered significant performance gains and reductions in fuel burn and carbon emissions, but the Rolls-Royce engines encountered a string of entry into service and ongoing durability issues that saw many 787s grounded for extended periods.

Passenger comfort was assured on the 787 by its higher cabin humidity, lower effective cabin altitude, larger windows with electronic dimming, improved air filtration and mood lighting. Combined with a quieter cabin, these features left passengers feeling fresher and less jet-lagged after a long-haul flight than on other widebodies.

Due to its lower operating costs and passenger appeal, the 787 turned long thin routes into viable markets with 787-9 launch customer Air New Zealand using it to good effect on services from Auckland to New York, Chicago and Los Angeles. US carriers such as United, Delta and American use 787s on transpacific flights to Australia while Qantas launched nonstop flights from Perth to London, Rome and Paris with the 787.

With more than 1,200 delivered by the end of 2025 and a further 1,026 on order, the 787 appeared at the right time to capitalise on the return of point-to-point flying and its flexibility in cabin layouts has allowed airlines to mix and match capacity to meet market demands. It has had some troubled

times, but with those in the past its path towards iconic status is well under way.

▲ **British Airways operates all three 787 variants.** British Airways

Short-haul star

One of the many lessons the COVID-19 pandemic taught us was the need to keep remote communities connected and in many parts of the world that means by air travel. Smaller jet or turboprop aircraft are often the only way to keep these connections open while these same aircraft are the backbone of regional passenger services in North America, Australia, Europe and Asia.

Brazilian aerospace company Embraer specialises in designing and manufacturing passenger jets with up to 150 seats, and about every ten seconds (on average) an Embraer aircraft takes off somewhere in the world. Since its start in 1969, the company has delivered more than 9,000 aircraft and today is a leader in passenger jets, business jets and military transport aircraft.

The short-haul or regional segment is undergoing significant change with many airlines relying on fleets of ageing, gas-guzzling jets or older turboprops that need to be replaced. There is also growing demand for regional aircraft in many developing nations where growing disposable incomes are driving the thirst for travel, often to second and third-tier cities.

New-generation aircraft like Embraer's E2 family and the Airbus A220 are setting new standards for smaller jet aircraft with cabins resembling modern narrowbody jets, increased performance and the ability to operate to airports that cannot handle 737 or A320 aircraft. On April 24, 2018, the first Embraer E190-E2 entered service with Norwegian regional airline Widerøe while the larger E195-E2 entered service with Brazil's Azul in October 2019.

◀ **Embraer's E190-E2 entered service with Widerøe.** Embraer

In January 2026, there were nearly 200 E2 aircraft in service, with the three largest operators being Porter Airlines (Canada), Azul (Brazil) and KLM Cityhopper (Netherlands). Embraer also had a backlog of more than 300 E2 jets with significant orders from Avelo (US), LATAM (Chile), SAS Link (Sweden) and All Nippon Airways (Japan).

Embraer introduced the E2 in 2013 in response to increasing demand for more fuel-efficient regional aircraft with lower operating costs and reduced environmental impact. The new aircraft was built around new technologies, such as the Pratt & Whitney GTF (geared turbofan engine), new flight systems and improved aerodynamics.

Compared with the E1 family the new-generation E2 has a redesigned wing with a higher aspect ratio enhancing cruise performance and efficiency, an updated flight deck featuring full fly-by-wire controls and advanced avionics that reduce pilot workload.

◀ **Etihad uses 787s on medium and long-haul routes.** Boeing

▲ **LCC Scoot is launching new routes with the E190-E2.** Embraer

▶ **Embraer manufactures its aircraft in Brazil.** Embraer

In a single-class layout, the E195-E2 can carry 146 passengers in a 2-2 configuration with a 3,000nm range, 41,000ft ceiling and a maximum cruise speed of 0.82 Mach. Its runway performance allows the E2 to operate at airports previously only serviced by turboprops and offers a cost-effective bridge between traditional regional jets and larger narrowbodies.

Its place in this section is based on the success it is having in opening up new markets in Asia, Australia, the Americas, Europe and Africa with a cleaner and more sustainable jet aircraft. Only time will tell if it becomes an airliner icon over the next decade, but the signs are promising.

▶ **Lower seat and operating costs are strengths of E2 jets.** Embraer